IMAGES
of America

OLD BUTLER

For Anna Dugger, our friend whose zeal and desire to preserve the history of Old Butler inspired us in our work.

IMAGES
of America

OLD BUTLER

Michael and Lanette Depew
with the Butler Museum

ARCADIA
PUBLISHING

ISBN 978-0-7385-4171-6

Published by Arcadia Publishing
Charleston, South Carolina

Printed in the United States of America

Library of Congress Catalog Card Number: 2005927756

For all general information contact Arcadia Publishing at:
Telephone 843-853-2070
Fax 843-853-0044
E-mail sales@arcadiapublishing.com
For customer service and orders:
Toll-Free 1-888-313-2665

Visit us on the Internet at www.arcadiapublishing.com

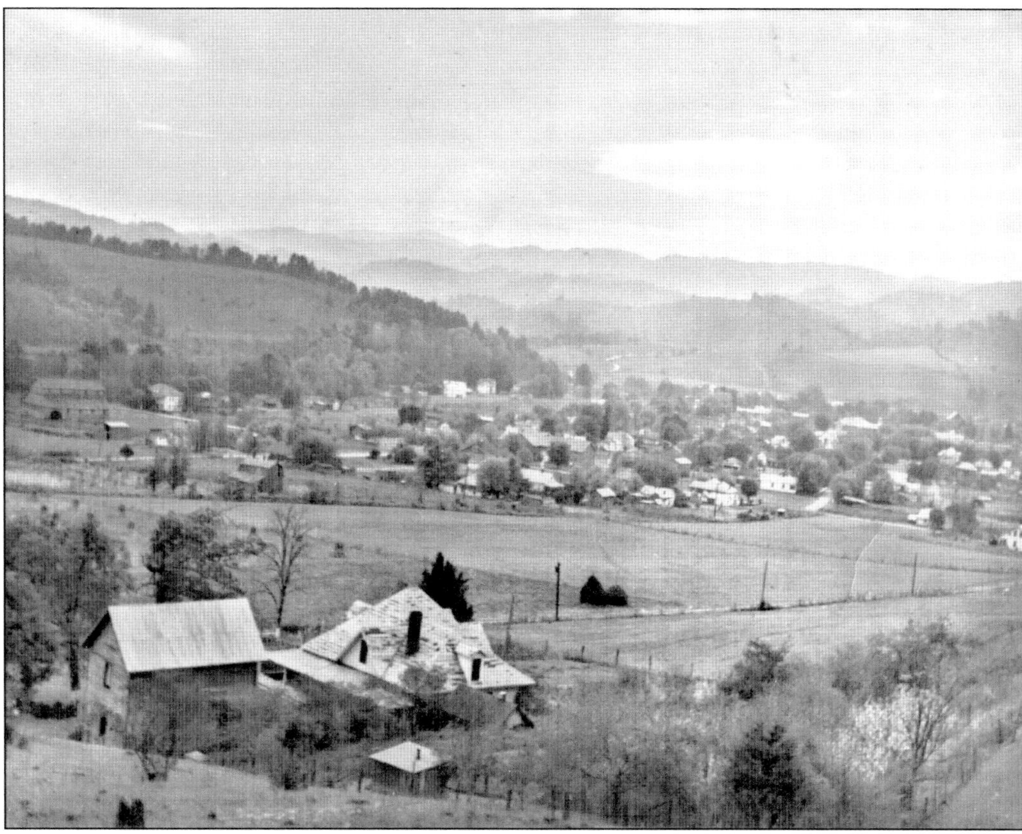

This photograph of Old Butler was taken in 1947. (Photograph courtesy of George Walker.)

Contents

ACKNOWLEDGMENTS

A very special thank-you goes to the number of people who guided us in our search for this collection's photographs. The history of Old Butler is intriguing, and it is our desire that with these images we can contribute to the preservation of its memory.

We would also like to thank our children, Lydia, Priscilla, Leo, Isaac, and John, for their patience and understanding while we put this work together.

Finally, we wish to extend our heartfelt appreciation to the Butler Museum and the Butler Watauga Valley Heritage Association for allowing us access to their enormous collection of photographs and documents and for their unfailing efforts to assist us in the completion of this project.

Two books which contributed to our research on Old Butler that we would like to recognize are Russ Calhoun Sr.'s *Lost Heritage: The People of Old Butler, Tennessee, and the Watauga Valley* (Johnson City, Tennessee: The Overmountain Press, 1998), and Dan Crowe's *Old Butler and Watauga Academy* (Johnson City: self-published by the author, 1983).

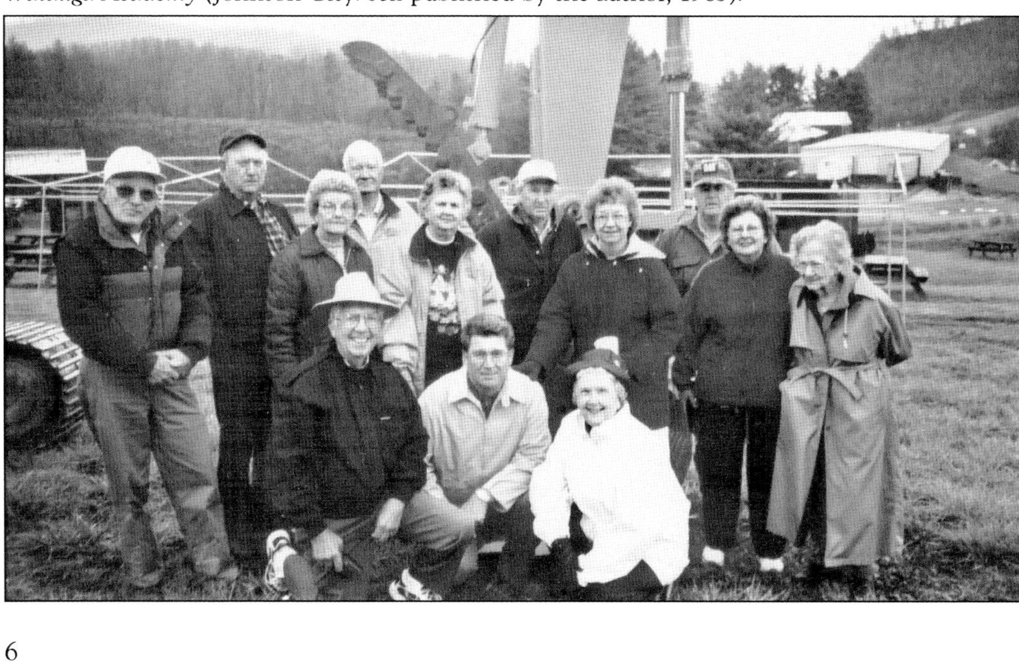

INTRODUCTION

In 1820, Ezekiel "Zeke" Smith built a gristmill on the bank of Roan Creek in what is now Johnson County, Tennessee. The community grew quickly, and by 1858, there were 155 families. Following the Civil War, the community, known for many years as Smith's Mill, was renamed Butler in honor of Col. Roderick Random Butler of the 13th Tennessee Volunteer Cavalry of the Union.

Even before the turn of the 20th century, Butler had begun to thrive with its farming, milling, and lumber industries along the banks of the Roan Creek and the Watauga River. The arrival of the railroad to the area not only created a consistent method of transporting goods to surrounding communities but also provided opportunities for the young men of the region to find gainful employment outside the Watauga Valley.

Watauga Academy, which started as a private school, maintained a reputation for academic excellence and attracted students from all over the region. James H. Smith not only led the school to its high rate of success but also helped to guide the community of Butler into the 20th century.

The city of Butler continued to thrive until the 1940s, when it was asked to make the ultimate sacrifice by the federal government and the Tennessee Valley Authority (TVA). The Watauga Dam was built, and by 1948, the remains of the city were laid to rest at the bottom of Watauga Lake.

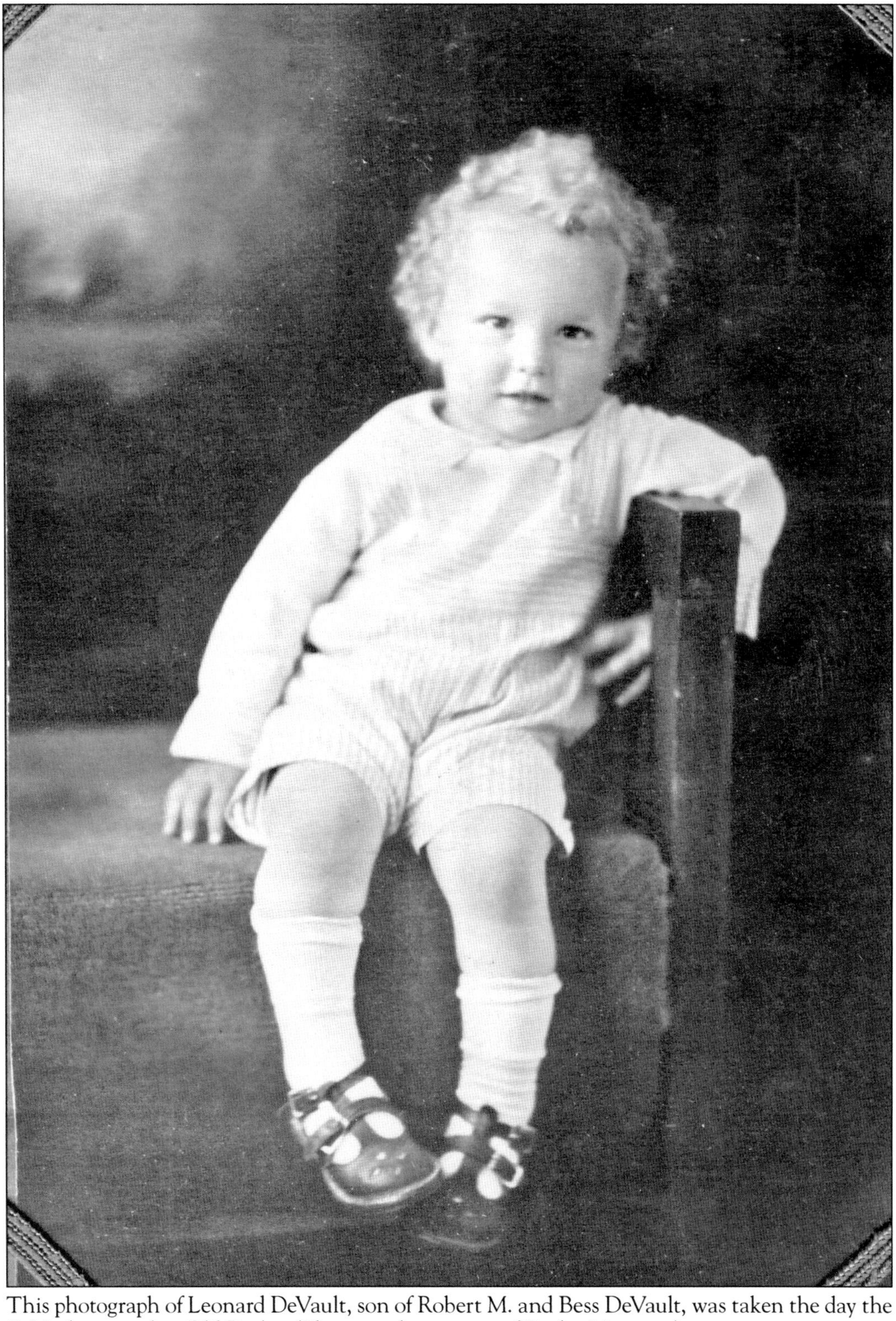

This photograph of Leonard DeVault, son of Robert M. and Bess DeVault, was taken the day the DeVaults moved to Old Butler. (Photograph courtesy of Butler Museum.)

One

A Town Develops

In 1858, merchants Thomas and Columbus Coffee established the Coffee Brothers' Mercantile, a successful general store near Roan Creek. During the Civil War, the Coffee brothers, who were Confederate supporters, left the settlement and fought for the South. When the war was over, the Coffee brothers returned to the community, re-established their business, and hired Finley P. Curtis to manage the store. In 1876, the Coffee brothers sold their store to Finley P. Curtis and J. D. Farthing. Curtis and Farthing built a livery stable and boarding houses that helped stimulate commerce in the region. James H. Smith, professor of Aenon Seminary (which later became Holly Spring College), not only directed the school but was also intricately involved in the building of the community itself. Smith even helped by surveying and laying out plans for the streets and sidewalks of the growing town. The entrepreneurs' fortitude, the ideal farming terrain, and the abundance of resources contributed greatly to the early growth of Butler.

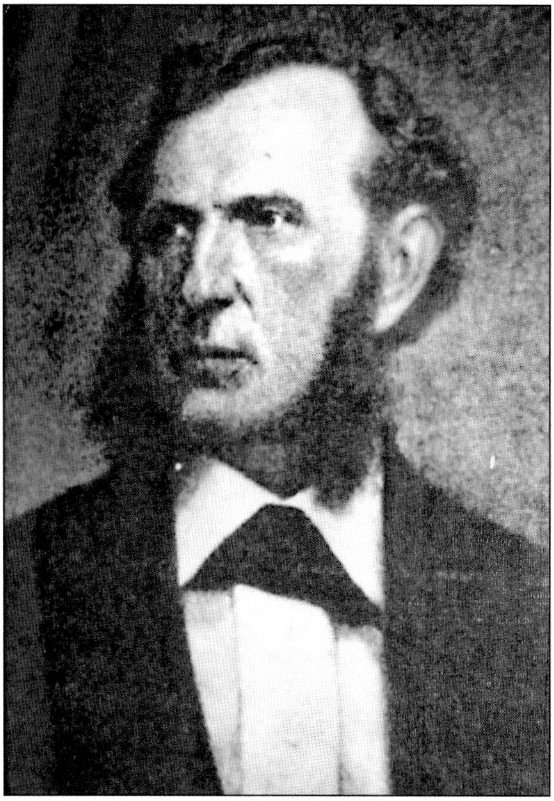

The city of Butler was named in honor of Col. Roderick Random Butler. Colonel Butler, who served with the 13th Union Volunteers during the Civil War, was later elected state representative. (Photograph courtesy of Butler Museum.)

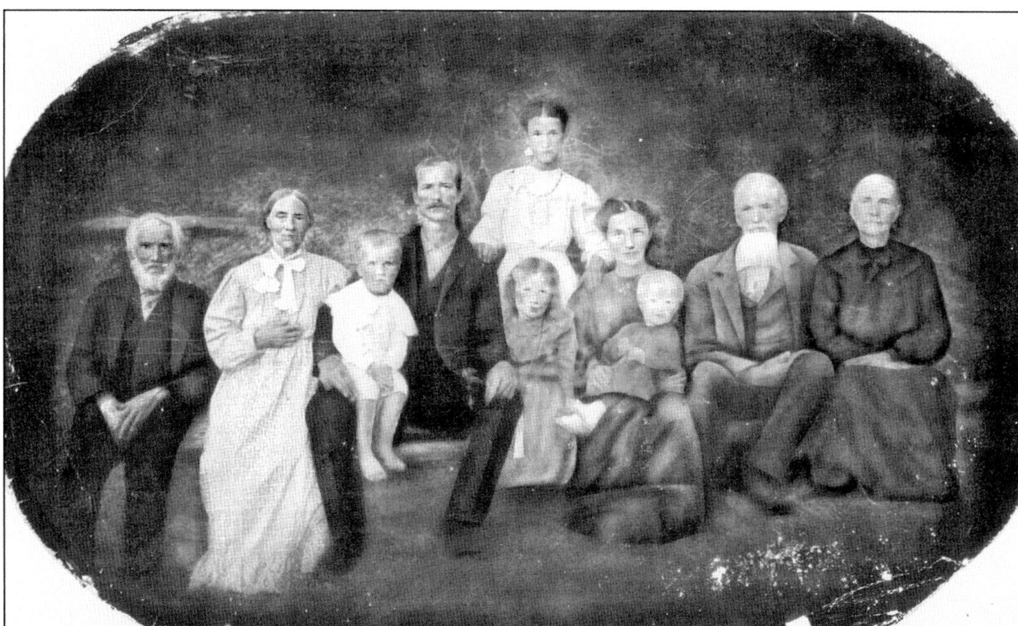

Pictured here is the Rainbolt family, who lived in the Little Milligan community. (Photograph courtesy of Butler Museum.)

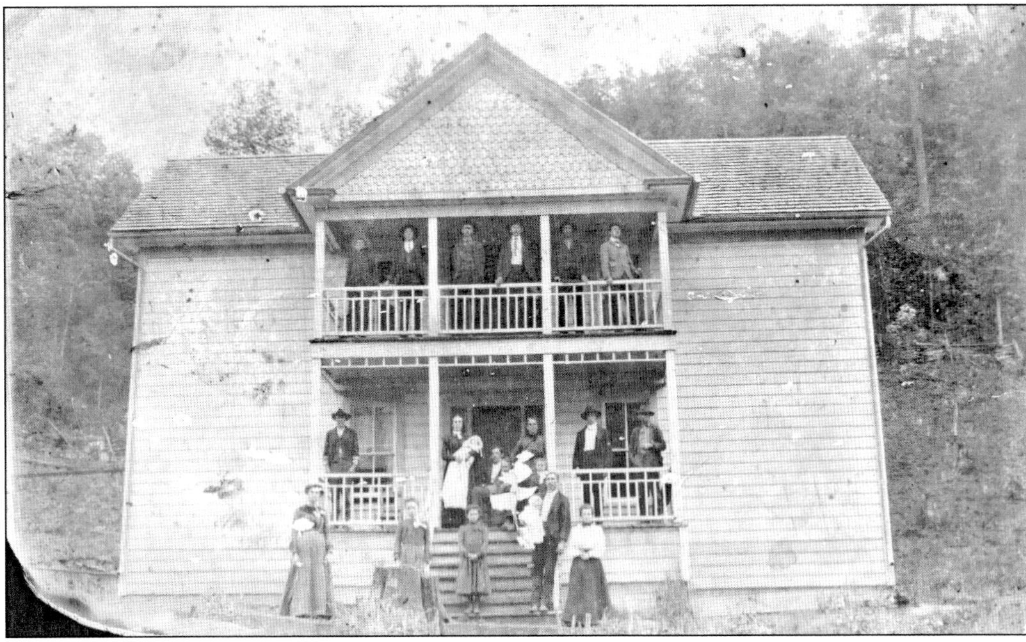

This photograph of Ollie Dugger (standing at left on the lower porch) with his family was taken in front of the Ollie Dugger house in the community of Buntontown. (Photograph courtesy of Butler Museum.)

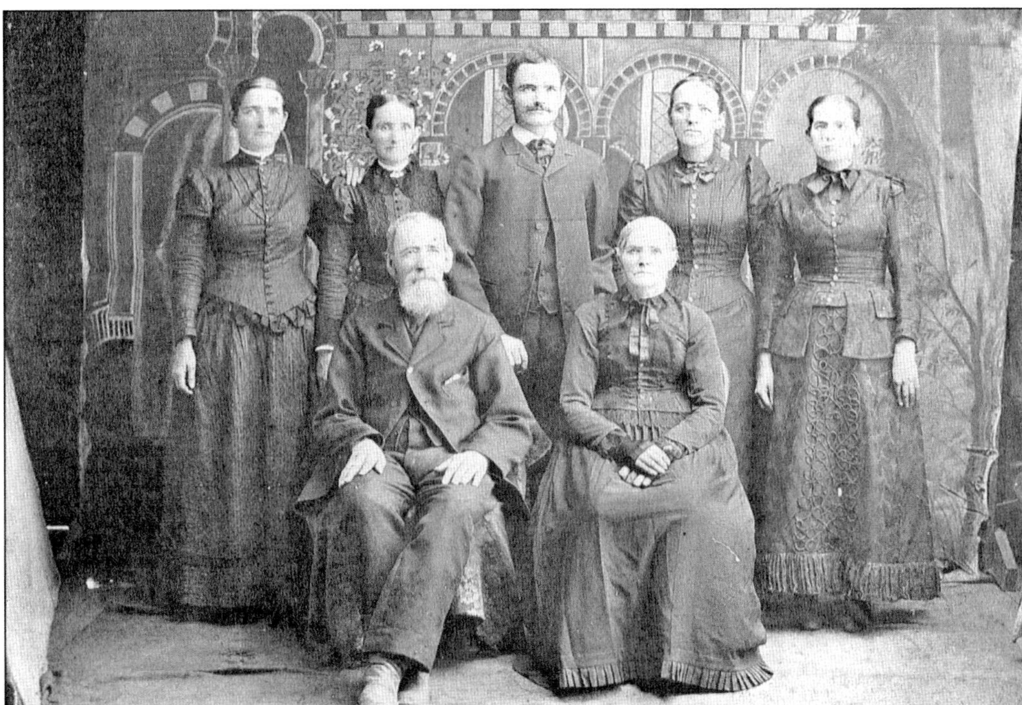

Pictured here are William Hardin Dugger and Nancy Cable Dugger with their children, standing from left to right, Elizabeth, Mary, Benjamin, Sarah, and Nancy. William and Nancy Dugger lived in the Sugar Grove community. (Photograph courtesy of Juanita Tester Wilson.)

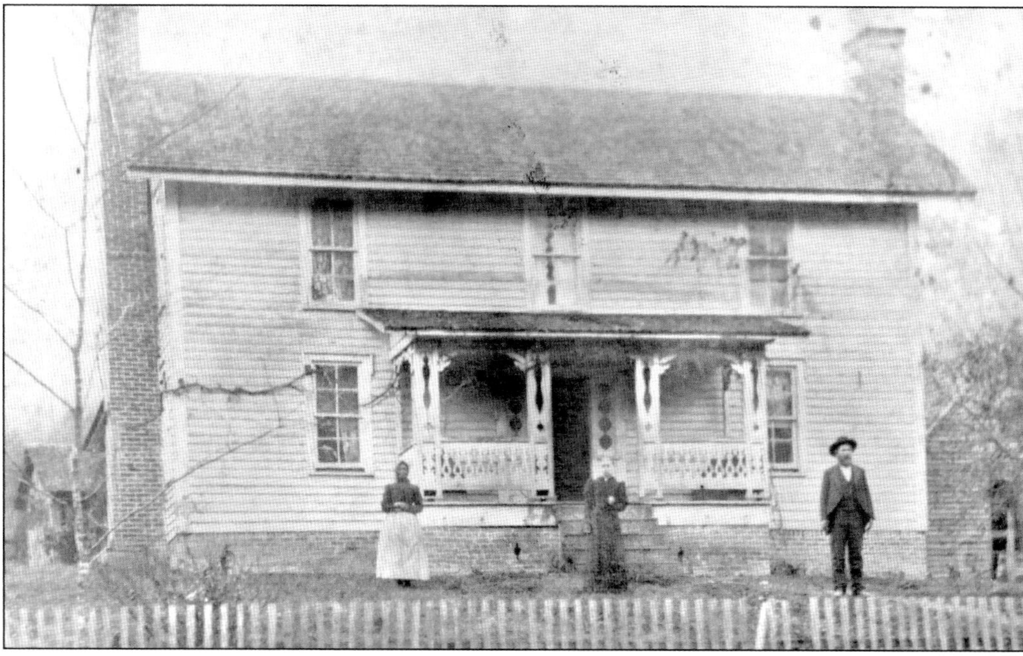

Daniel B. Baker (right) stands with his wife, Sarah Vaught Baker (center), in front of the house he built in 1870 in the Baker's Gap area. With Daniel and Sarah is their servant, Minerva Vaught (left), a wedding gift from Sarah's parents, John and Rebecca Vaught. Minerva served three generations of Bakers. (Photograph courtesy of Juanita Tester Wilson.)

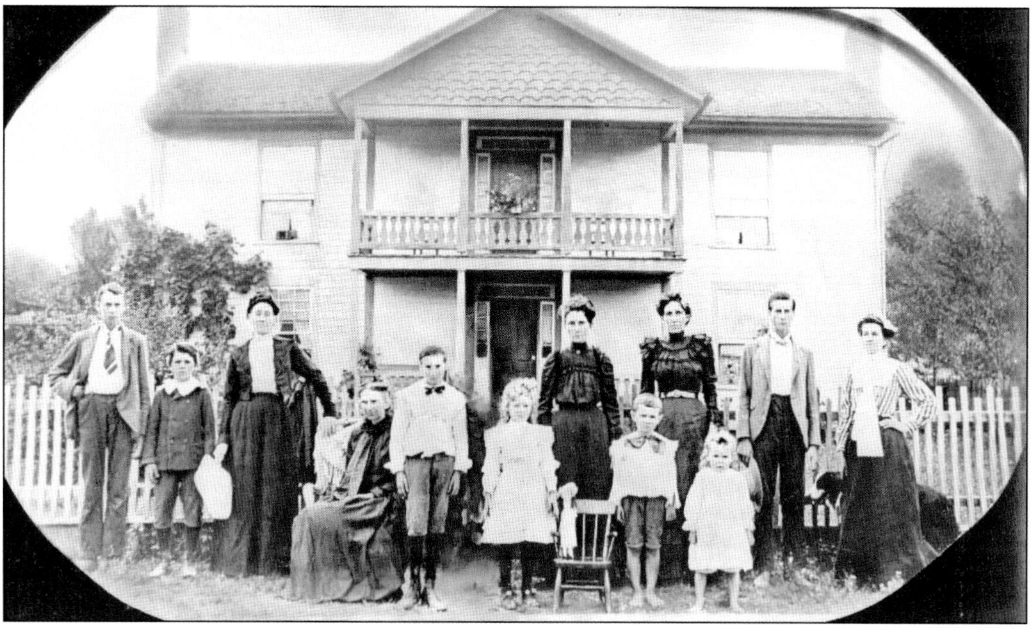

From left to right, George Thurston Walker, Wayne Fritts, Karen Walker Fritts, Mary Elizabeth Carriger Walker, Glenn Glasco Walker, Maude Goodwin (Weaver), Lucinda Walker, Paul Goodwin, Catherine Walker, Ruth Goodwin, Lee Goodwin, and Winnie Walker Goodwin pose in front of their home located near Doe Creek. The photograph was taken c. 1900. (Photograph courtesy of Butler Museum.)

Pictured here is Mary Elizabeth Carriger Walker. The Walkers and Carrigers were among the many prominent families in Butler. (Photograph courtesy of Mary Walker Ward.)

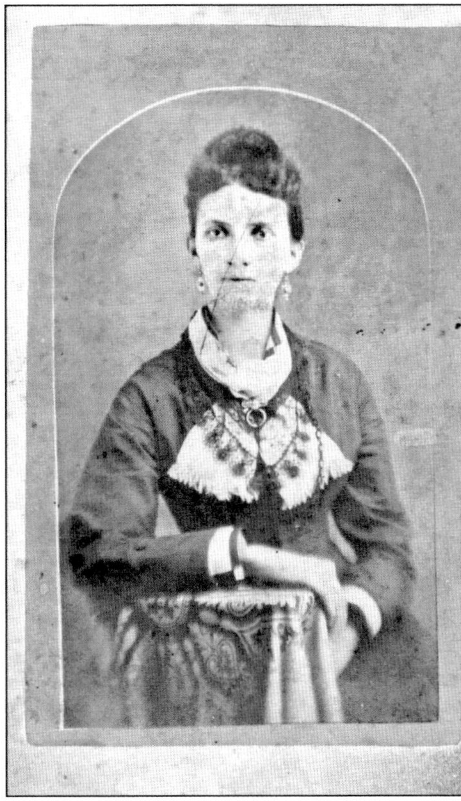

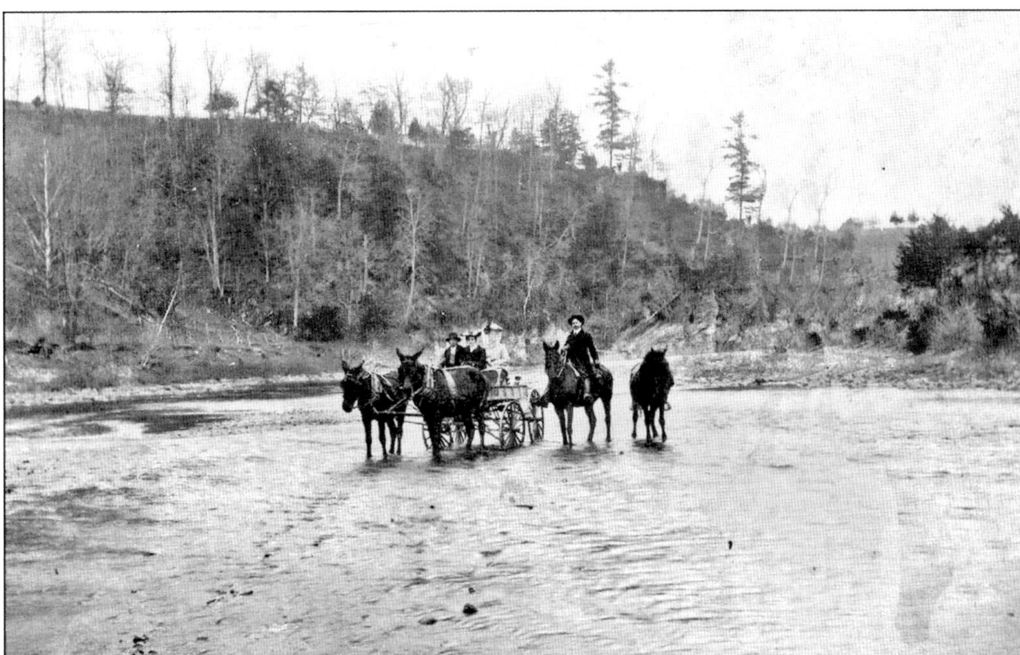

This photograph, taken during the early 1900s, shows residents of Butler fording the Watauga River. (Photograph courtesy of Butler Museum.)

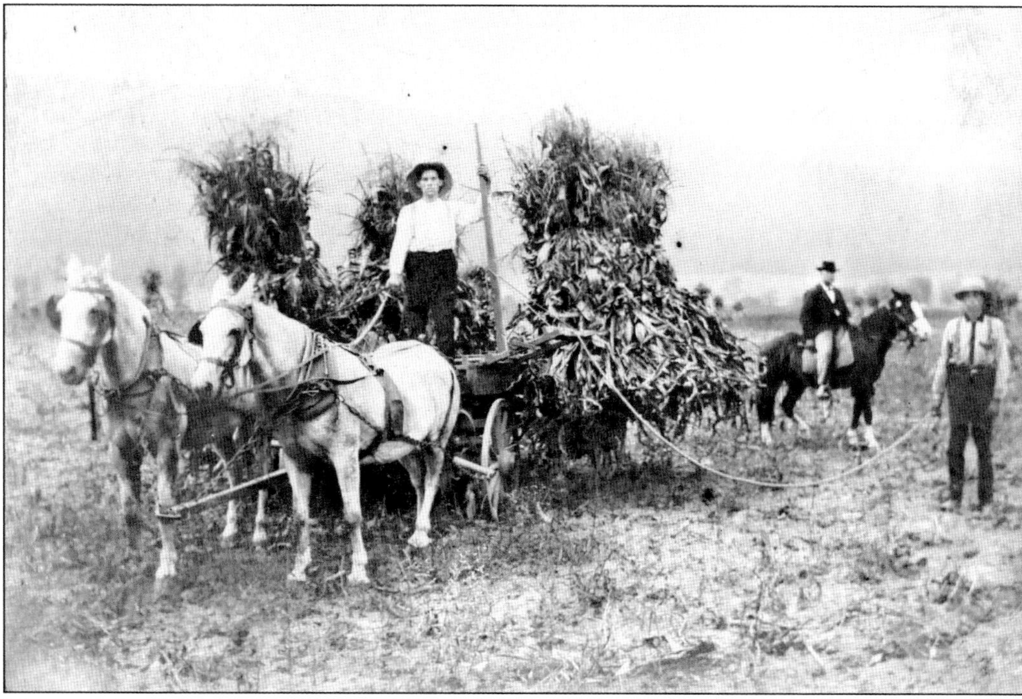

This corn-loading machine, also called a "fodder shuck," was owned by F. P. Curtis and J. D. Farthing. (Photograph courtesy of Butler Museum.)

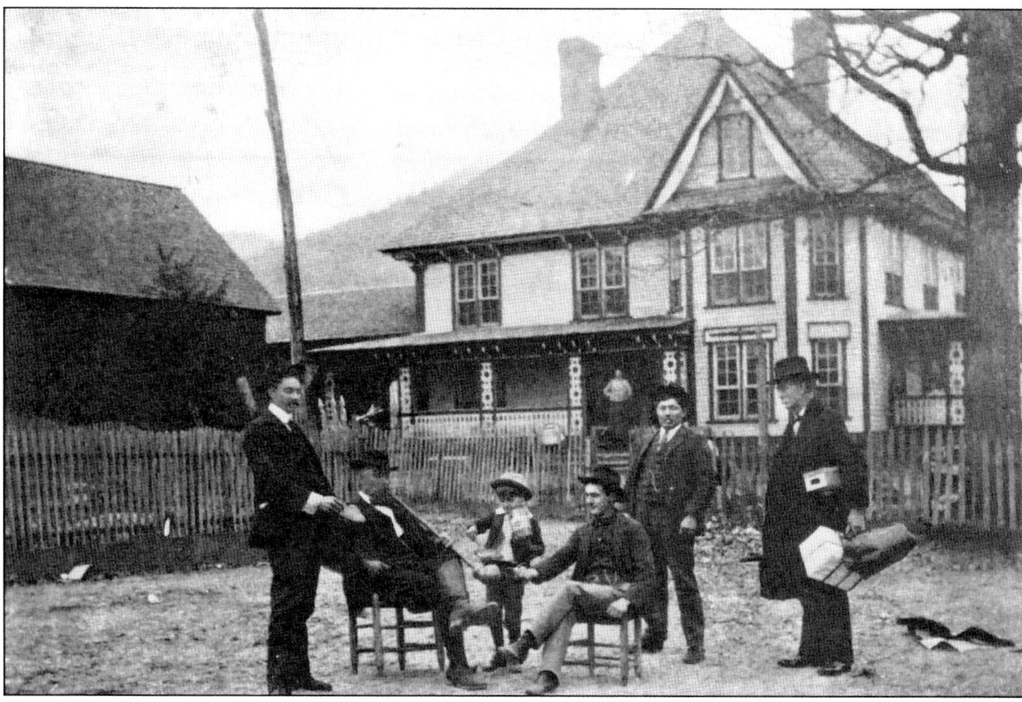

After purchasing the Coffee Brothers' store from Thomas and Columbus Coffee in 1876, F. P. Curtis and J. D. Farthing contributed greatly to education and growth in Butler. Pictured here is the Alabama House. J. D. Farthing is seated on the left. (Photograph courtesy of Butler Museum.)

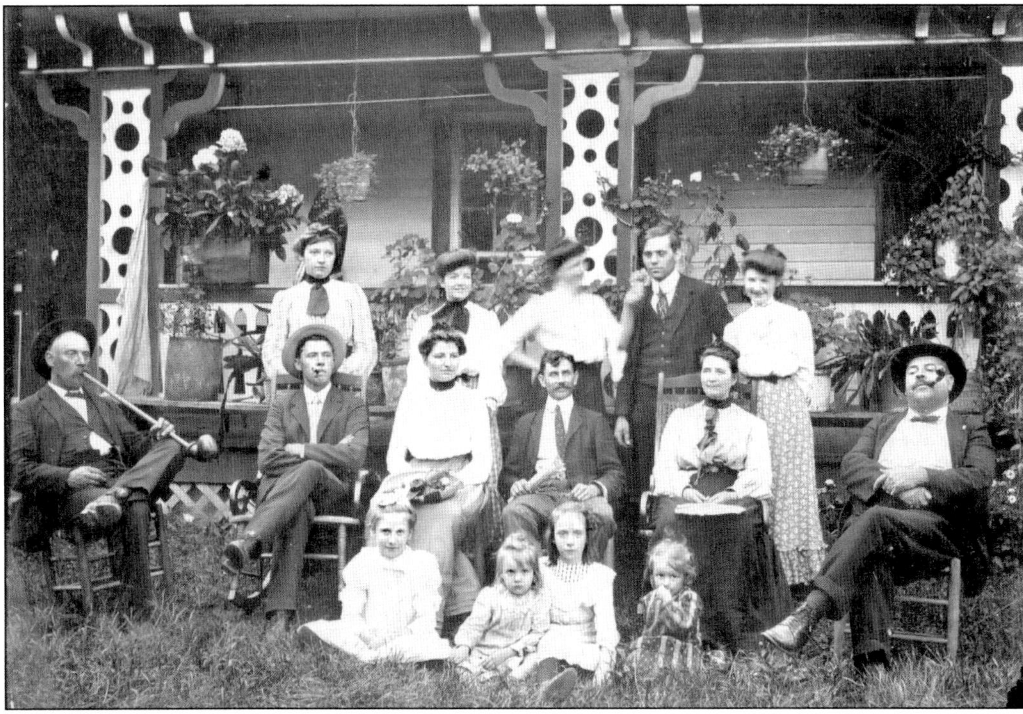

D. H. Farthing's home, later called the Alabama house, was a boarding house used by patrons of the Curtis-Farthing Livery Stable. D. H. Farthing is seated on the left. F. P. Curtis Sr. is seated on the right. (Photograph courtesy of Butler Museum.)

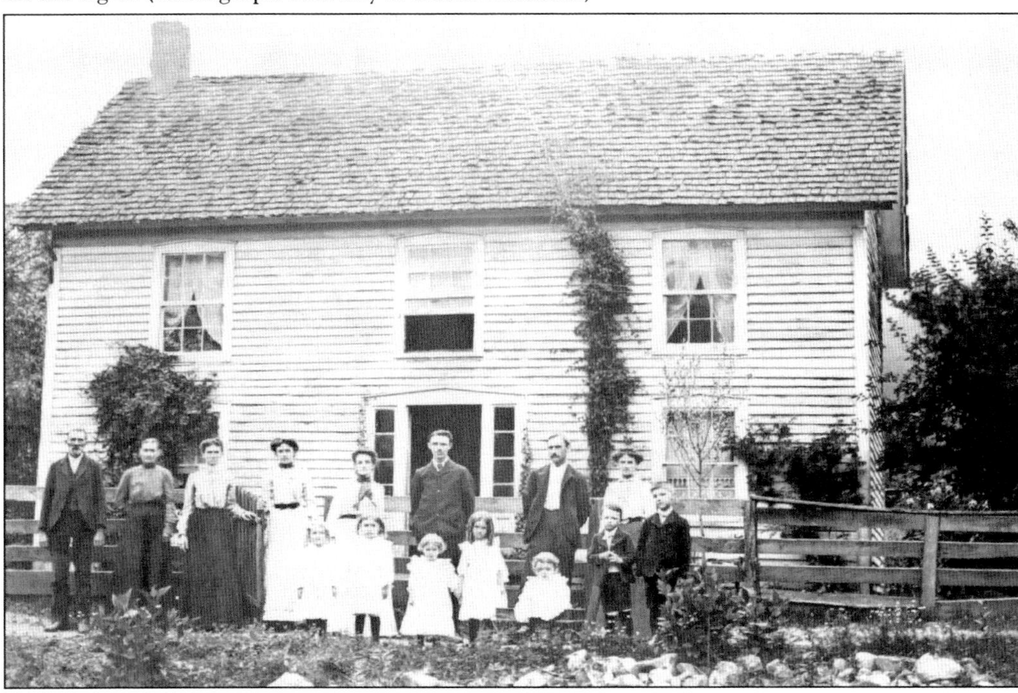

Four generations of the J. B. Fletcher family pose in front of their house in the Sink Valley community. (Photograph courtesy of Butler Museum.)

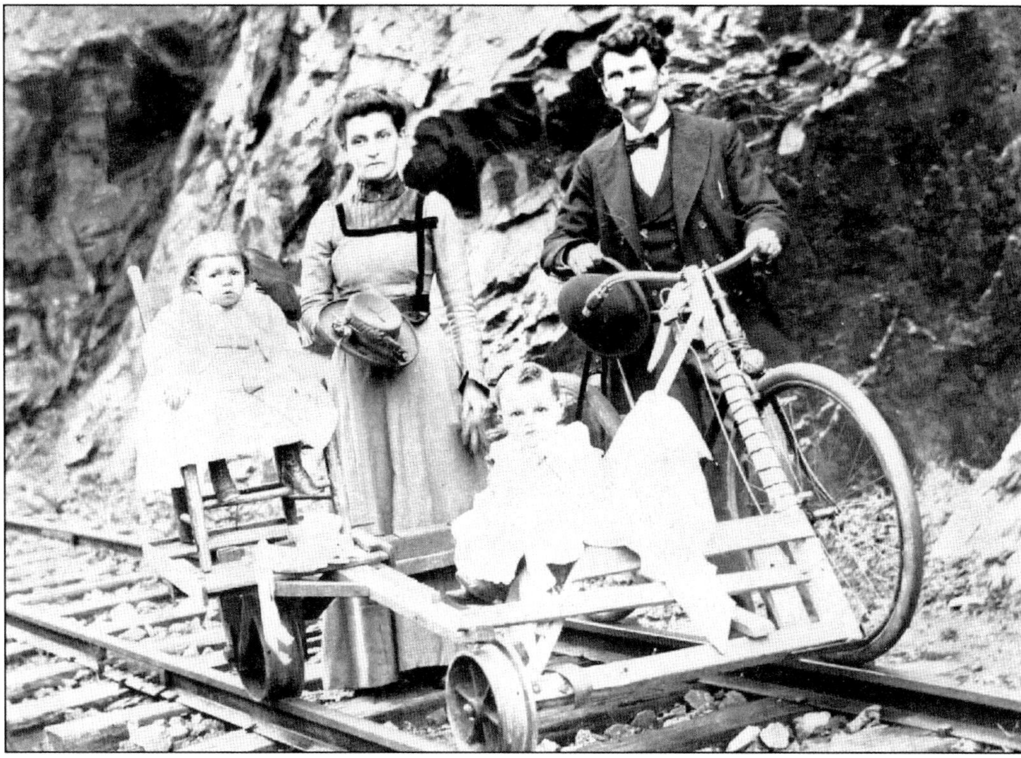

The Fletcher family is shown here riding up and down the rails on a makeshift railroad cart. (Photograph courtesy of Butler Museum.)

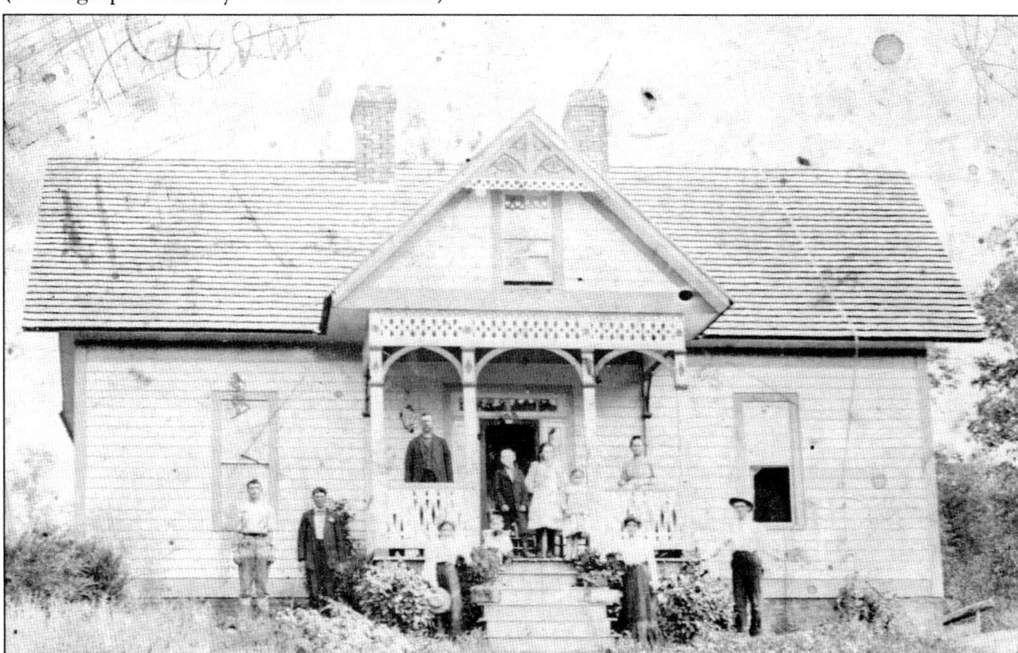

Pictured here are Andrew Hatley (on the porch, far left) and his family. (Photograph courtesy of Butler Museum.)

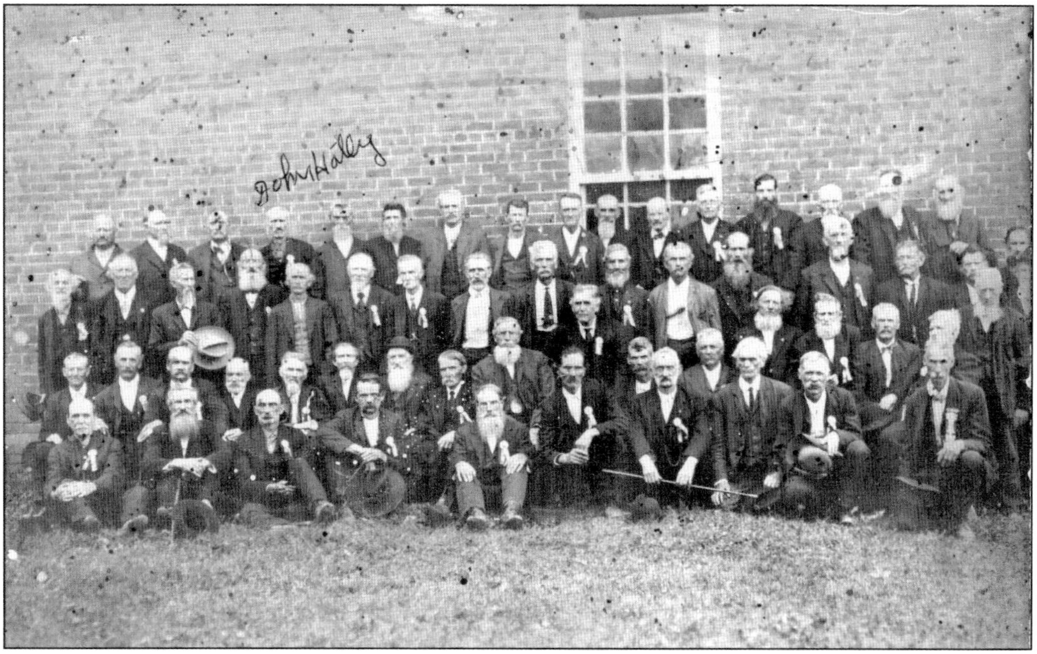

This photograph of a group of Butler men was taken *c.* 1900. The medals on the lapels suggest they belong to some type of organization. (Photograph courtesy of Butler Museum.)

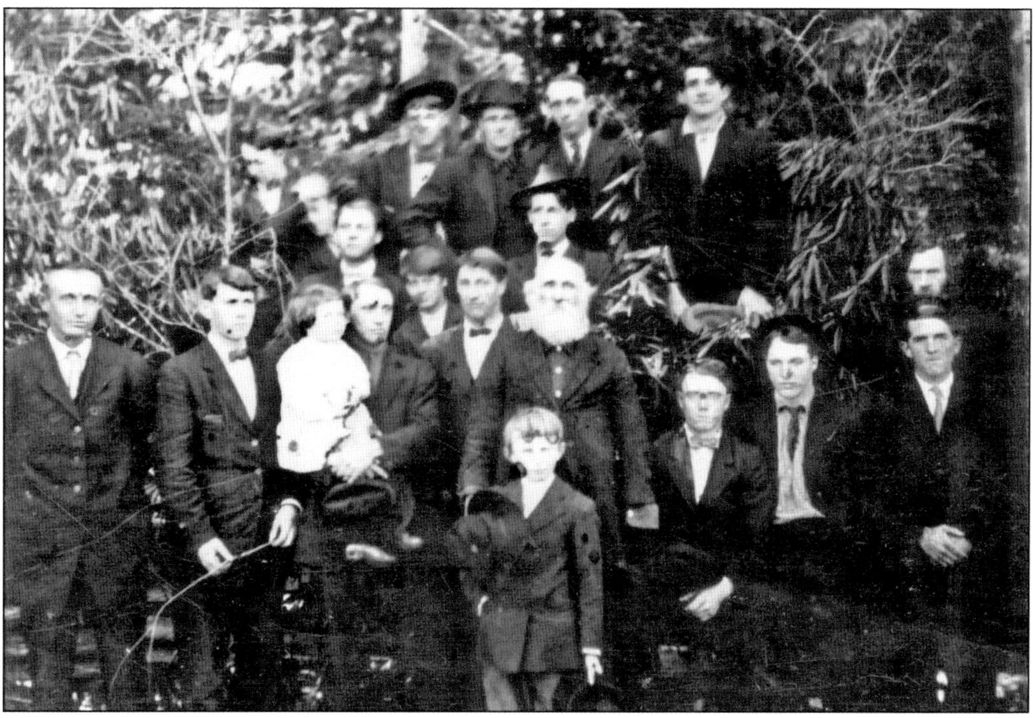

This photograph of the Farthing family includes the following: Clyde Perry, Jim Farthing, Jack Farthing, Bert Farthing, Bennie Farthing, Walter Cook, Robert Farthing, Carl Farthing, Carter Farthing, Raleigh Farthing, Bob Farthing, Fred Baird, and Ray Wilson. The Farthing family lived in the Sugar Grove community. (Photograph courtesy of Patsy Tadlock and Ermine Arwood.)

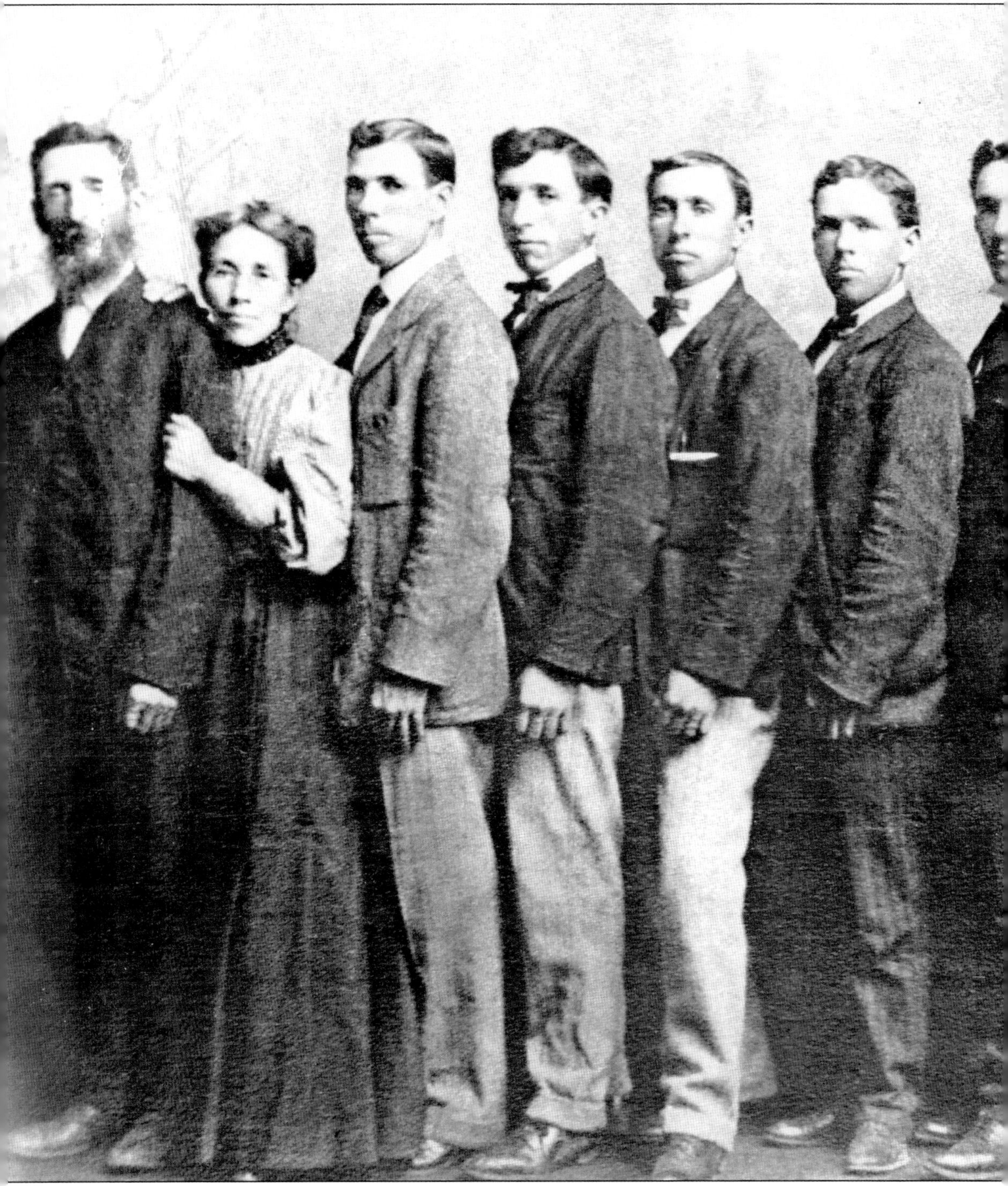

John (far left) and Sarah Maples Reece (second from left) are pictured here with their children, c. 1906. The children are, from left to right, James LaFayette, Wilson, John, Walter, Joseph Isaac, Amanda Catharine, Asa, Brazilla Carrol, Lemiel, Millard, Anna Elizabeth, Laura, and

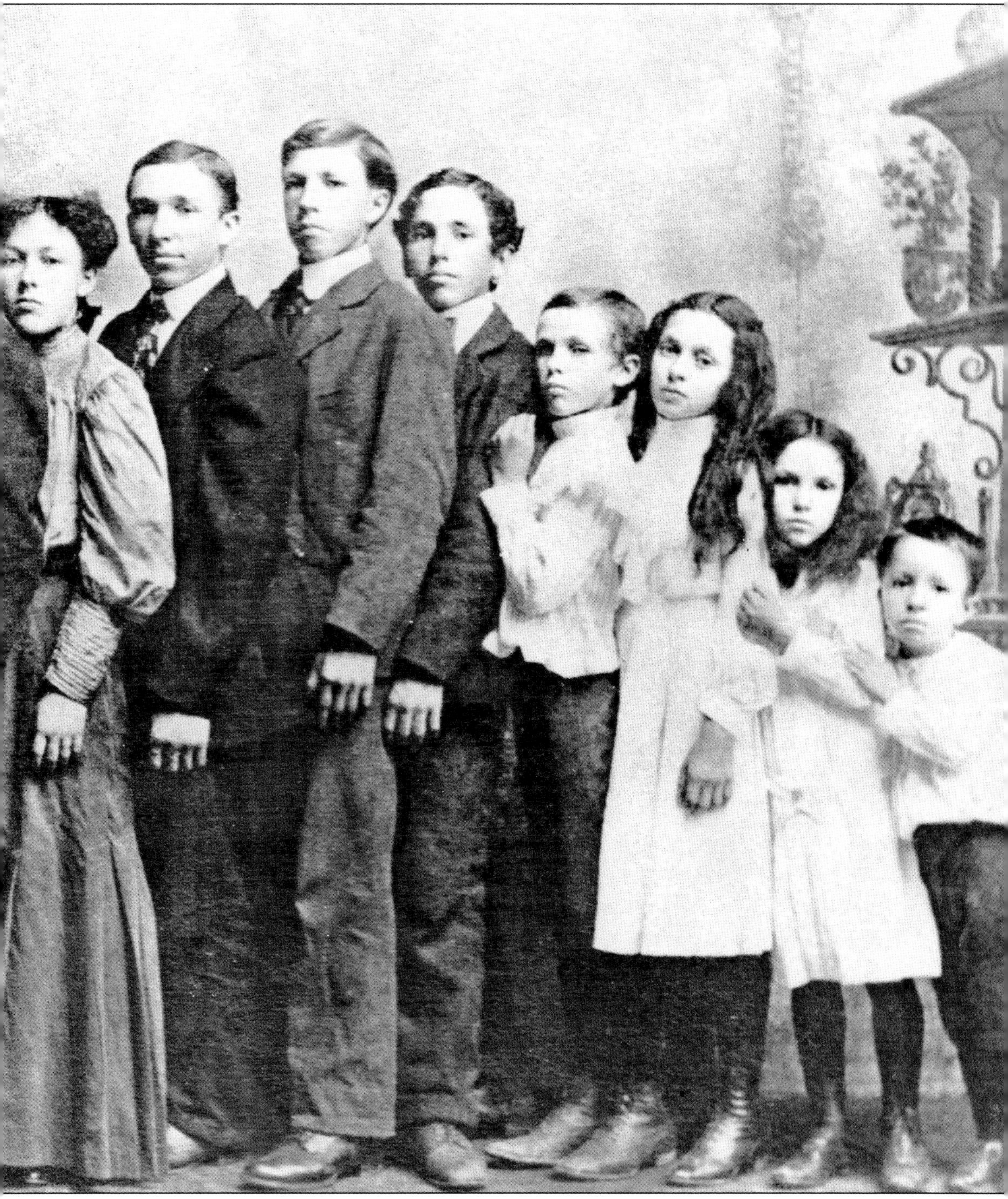

Raleigh Valentine. Fraher Norman, who was born between Laura and Raleigh, may have been deceased when this photograph was taken. (Photograph courtesy of Butler Museum.)

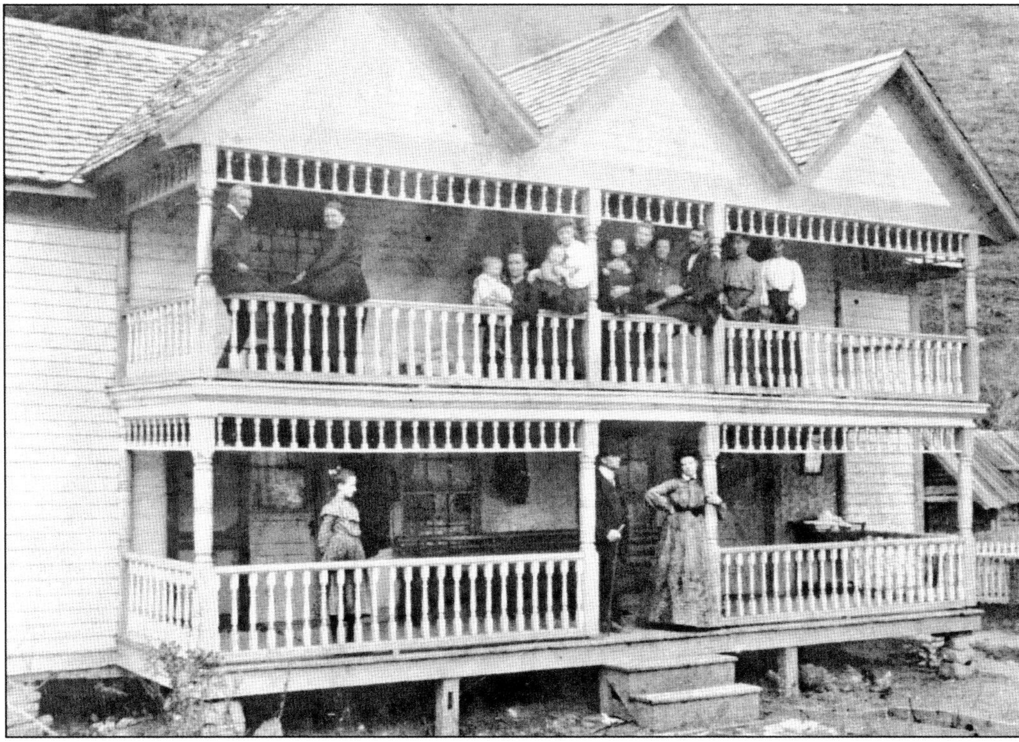

Pictured here is the McQueen family in 1915. L. L. McQueen not only owned the McQueen Lumber Company, but was also instrumental in the building of the railway from Horseshoe Bend to Butler in 1901 and to Mountain City in 1902. (Photograph courtesy of James D. Robinson.)

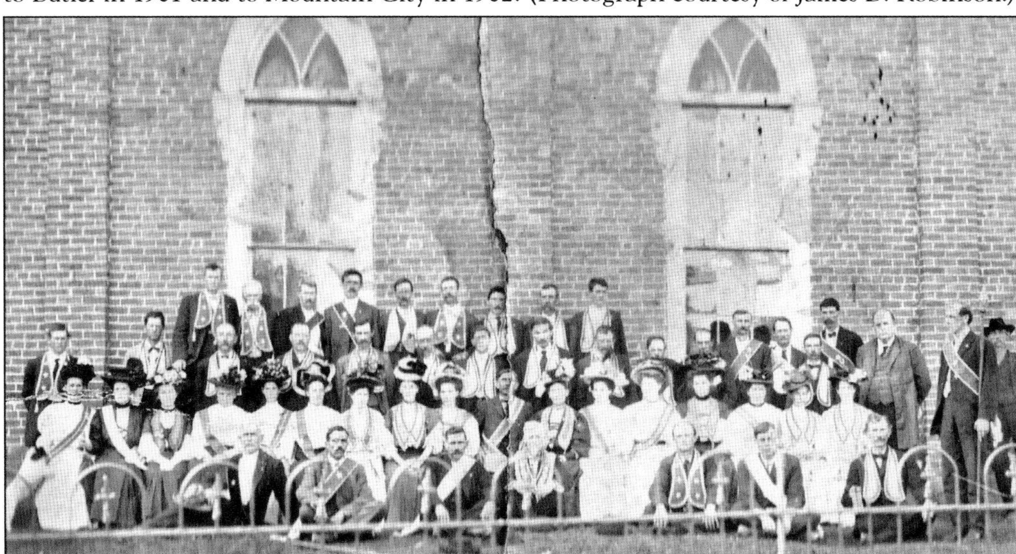

For more than 200 years, the chief aim of the Order of Odd Fellows and Rebekahs has been to aid in the welfare of its members. This photograph was taken near the Butler Baptist Church in the early 1900s and includes Ike Courtner, Houston Smith, Dr. Proffitt, ? Shoun, Squire Goodwin, and Preacher Todd. The organization met on the third floor of the Farmer's Union Store on Main Street in Butler. (Photograph courtesy of Butler Museum.)

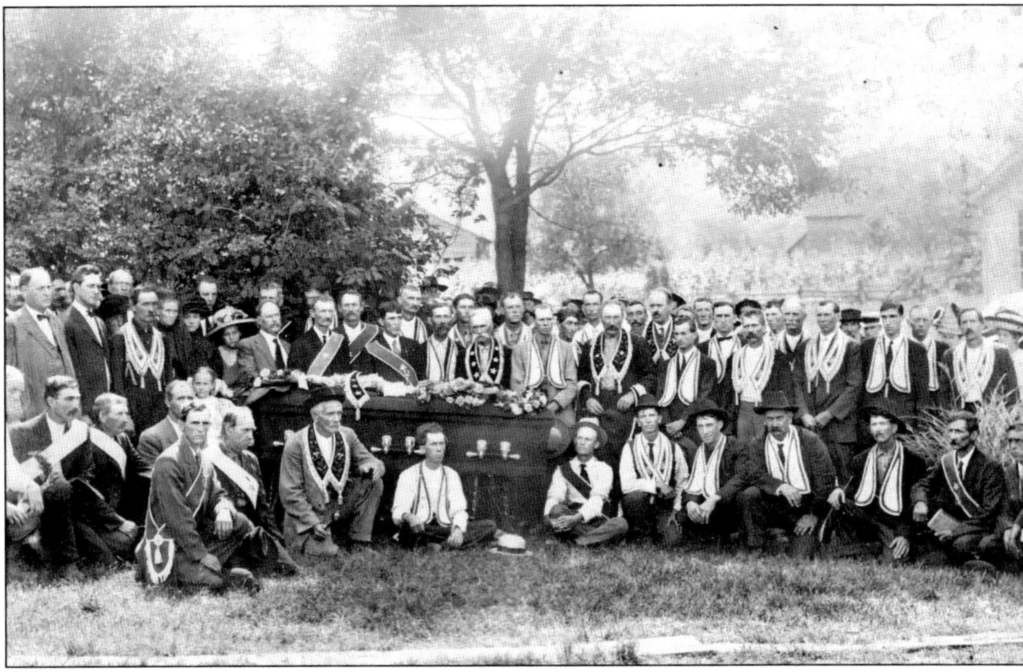

The Odd Fellows used a telegraphic cipher and key to make communication by telegraph between lodges less expensive. A single code word would convey a request or an answer regarding lodge business. This photograph, taken in the early 1900s, is of the funeral of an unidentified member of the Odd Fellows Lodge. (Photograph courtesy of Butler Museum.)

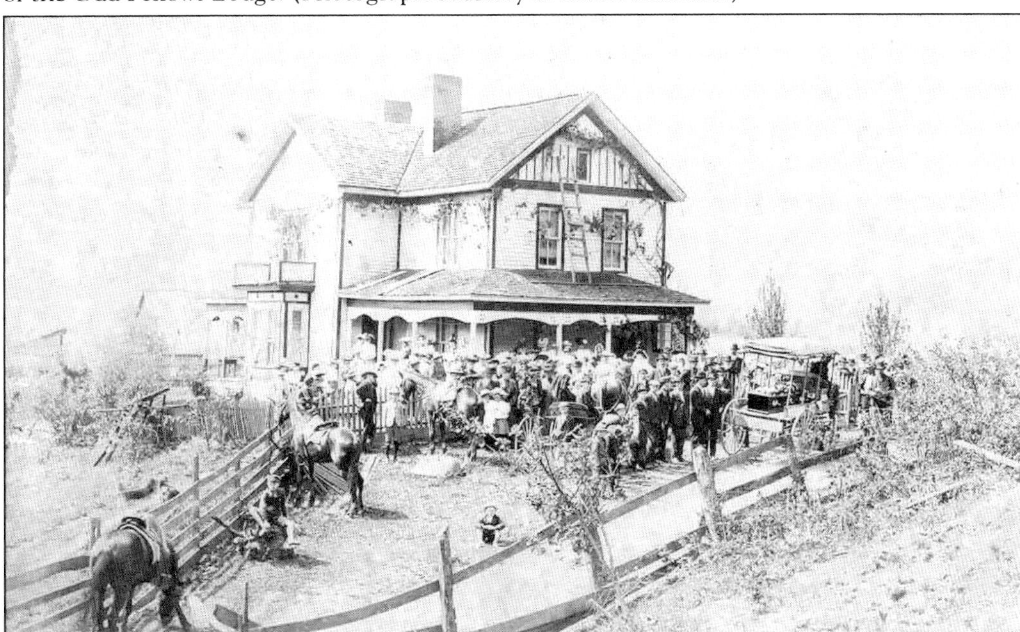

Family and neighbors attended the funeral of Preston Goodwin on May 16, 1907. Preston, who was only 15 years old when he died, was the son of Frank and Cindy Goodwin. The home in the background belonged to John Franklin Goodwin and Cindy Goodwin. (Photograph courtesy of Mary Lou Gross.)

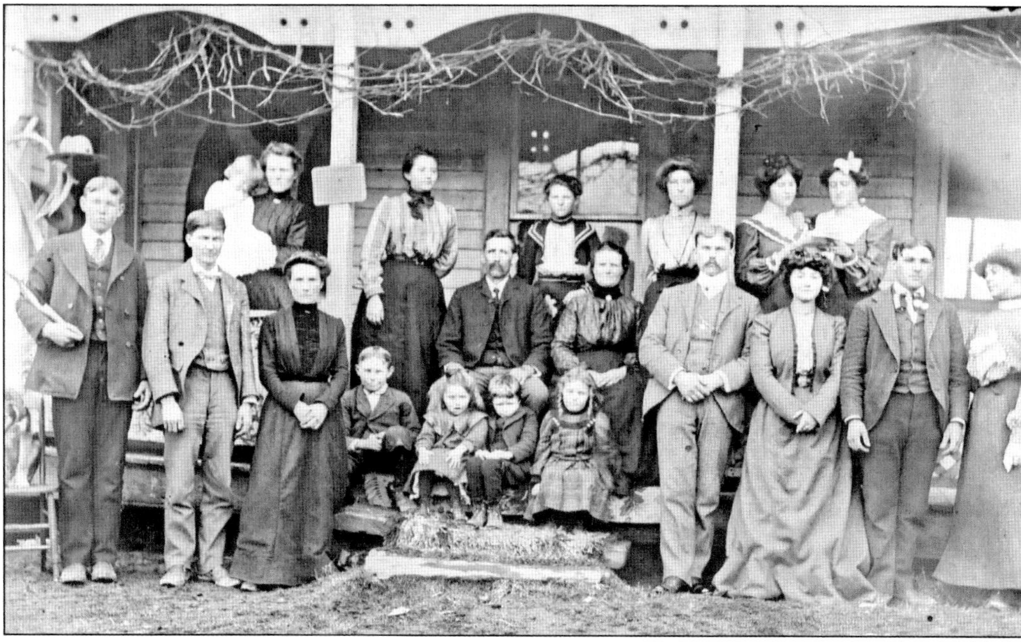

This photograph taken in 1902 depicts the wedding of Stacy and Amanda Goodwin. Pictured from left to right are (first row, standing) Austin Goodwin, Bob White, Mrs. Bob White, Stacy Goodwin, Amanda Moreland Goodwin, Will Goodwin, and Bess Shoupe; (children on the steps) Preston Goodwin, Luna Smith, Wylie Smith, and Stella Goodwin; (seated) Frank Goodwin and Lucinda Goodwin; (back row, standing) Doran Smith, Minnie Smith, Mrs. Alice Campbell, Mollie Lipford, Mamie Goodwin, Lena Barker Cotrell, and Ashley Smith. (Photograph courtesy of Mary Lou Gross.)

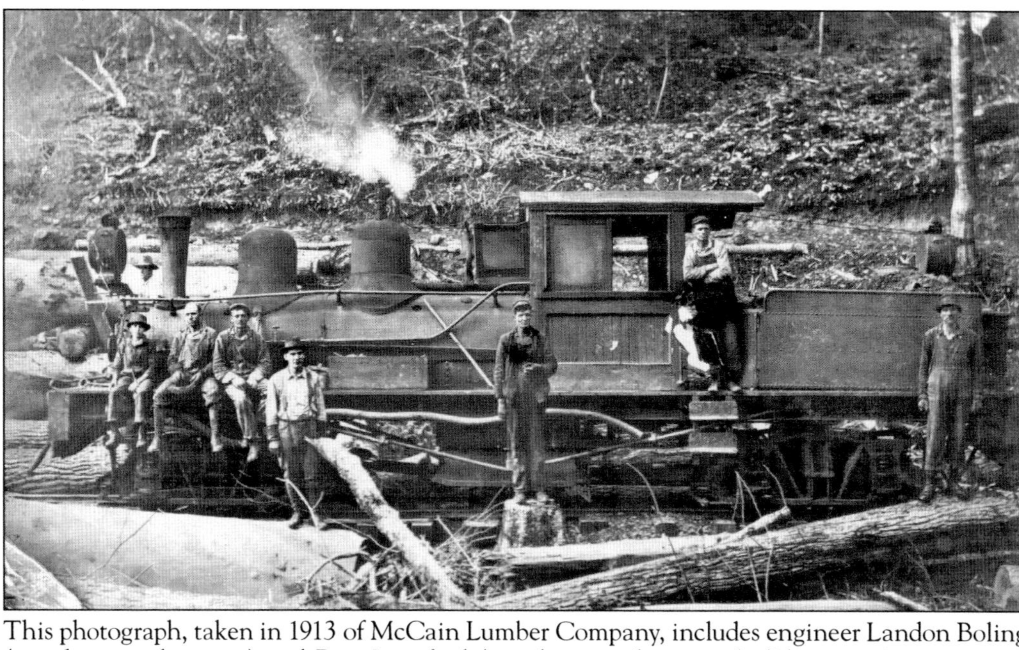

This photograph, taken in 1913 of McCain Lumber Company, includes engineer Landon Boling (standing on the train) and Bert Lunsford (standing on the stump). (Photograph courtesy of Butler Museum.)

After growing up in Butler, Tennessee, and graduating from Watauga Academy in 1911, B. Carroll Reece spent a distinguished career as representative for the first district of the State of Tennessee in the U.S. House of Representatives. In 1942, Mr. Reece honored the graduating class of Watauga Academy with a speech, which applauded the successes of the school through the years. (Photograph courtesy of Butler Museum.)

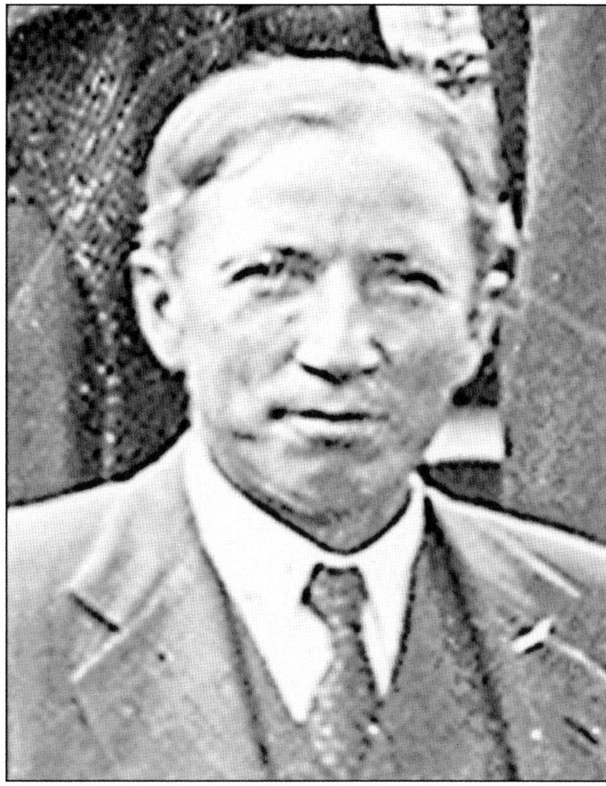

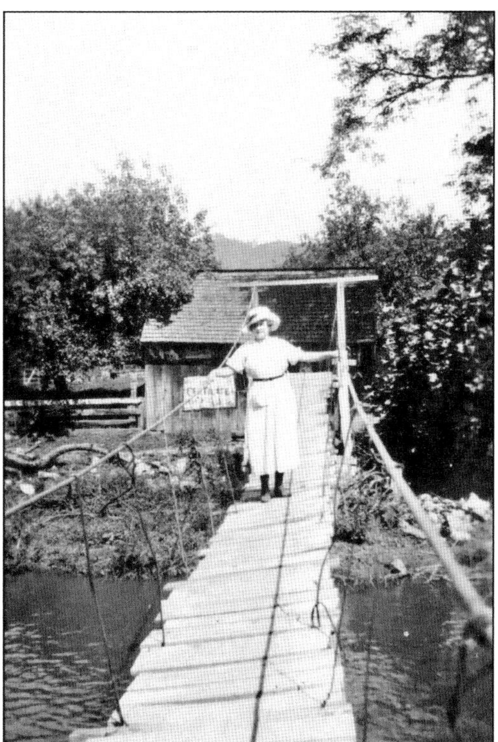

Mae Dugger Day stands on a swinging bridge that extended over Roan Creek. Notice the Whiting Lumber Company in the background. (Photograph courtesy of Mary Walker Ward.)

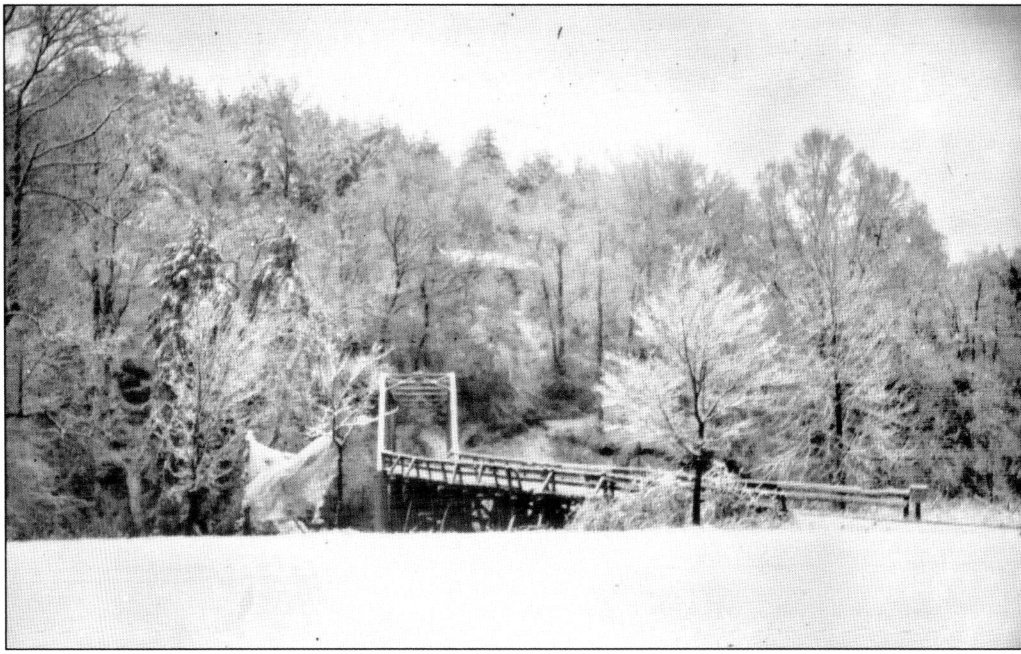

Pictured here is the Roan Creek Bridge. At one time, a large Native American community had resided in the Johnstown community and Butler Springs, located just beyond the Roan Creek Bridge. Many Native American artifacts have been found in that area. (Photograph courtesy of Butler Museum.)

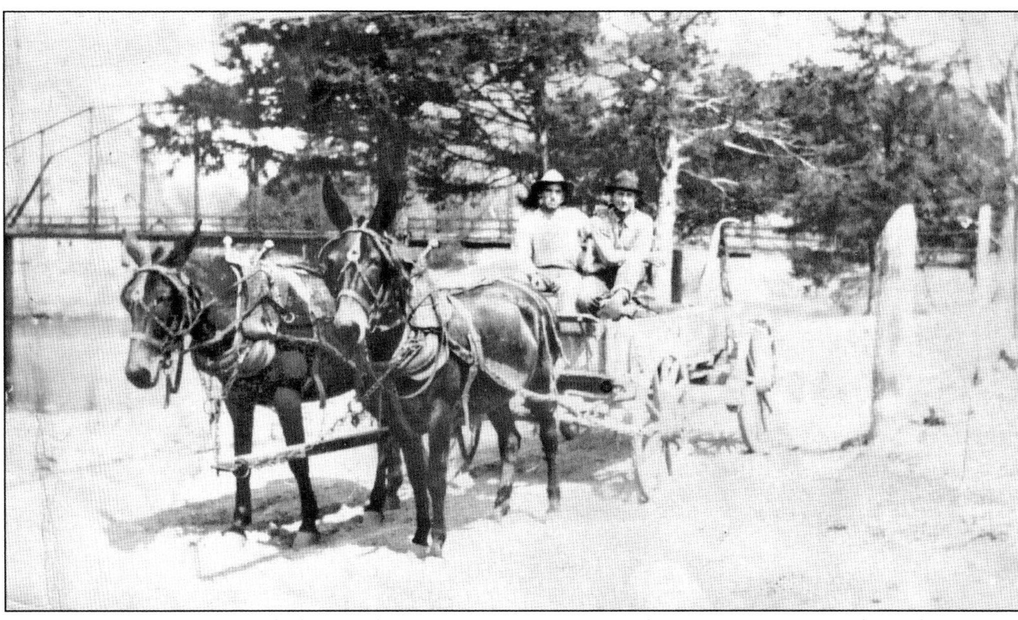

The Butler Bridge extended over the Watauga River, providing access into Butler. This image, with the bridge appearing in the background, was taken c. 1920. (Photograph courtesy of Butler Museum.)

Two

BUSINESSES AND ORGANIZATIONS

Farming, milling, and lumber industries contributed greatly to economic stability in the Watauga Valley. The fertile land provided rich farmland and a vast supply of lumber. Lumber companies included McCain Lumber Company, Whiting Lumber Company, and the A. H. McQueen Lumber Company, some of which developed to a point where they could afford to run their own narrow-gauge train lines. Glenn and George Walker formed the Walker Brothers Power and Light Company on the banks of Roan Creek. In this operation, flour and meal were ground by day, and electricity was generated for the residents of Butler from 6:00 p.m. until midnight. The Virginia & Southwestern Railway Company connected Butler with Mountain City, Hampton, Elizabethton, and even Western North Carolina in 1901, which assisted in the region's further growth. Due to these contributing factors, Butler quickly became a developing community boasting a depot, bank, electric-power company, several small businesses, a sewer system, and paved roads and sidewalks. Along with industrial growth, the community also developed professionally. A number of doctors practiced in the area, and individuals such as B. Carroll Reece and James D. Robinson took an active role in politics both locally and nationally.

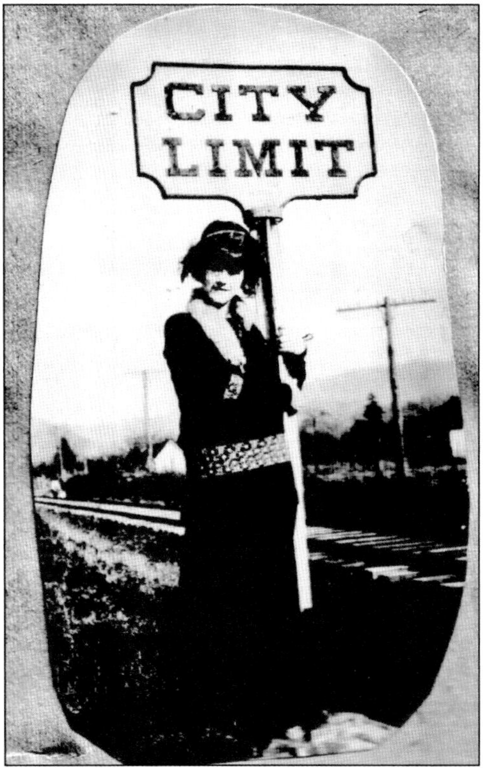

Aileen King poses near the city-limit sign of Butler in this photograph taken during the 1920s. (Photograph courtesy of Butler Museum.)

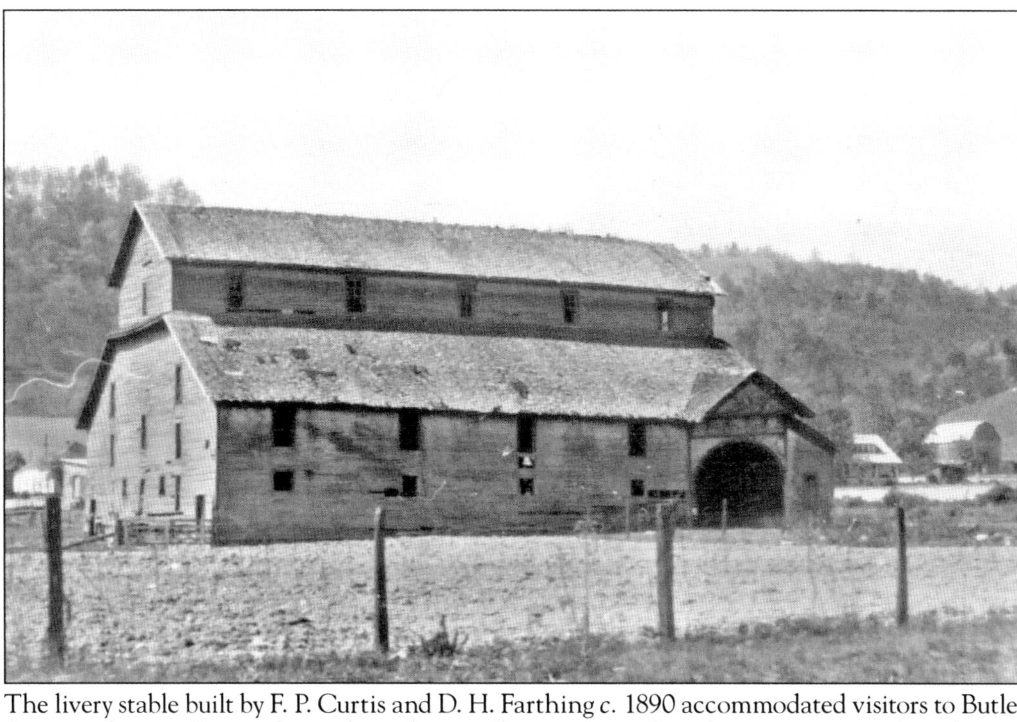

The livery stable built by F. P. Curtis and D. H. Farthing *c.* 1890 accommodated visitors to Butler for several years. Rooms located on the top floor were used as sleeping quarters for salesmen known as "drummers." (Photograph courtesy of Butler Museum.)

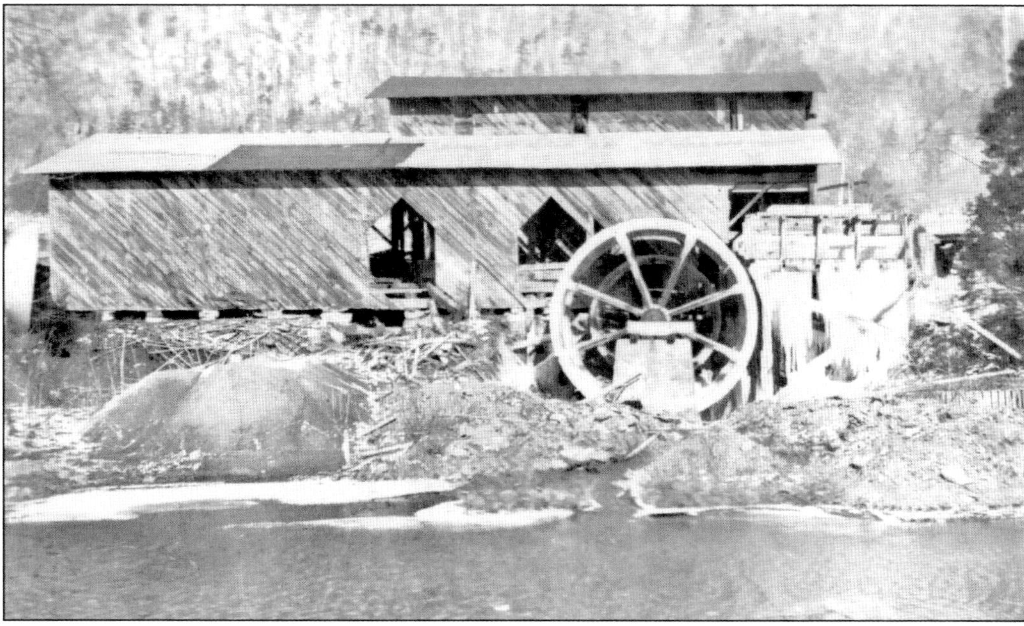

Located on Roan Creek, Whiting Lumber Company had a thriving lumber business that served the community for many years. Pictured here is the waterwheel, which powered the sawmill. (Photograph courtesy of George Walker.)

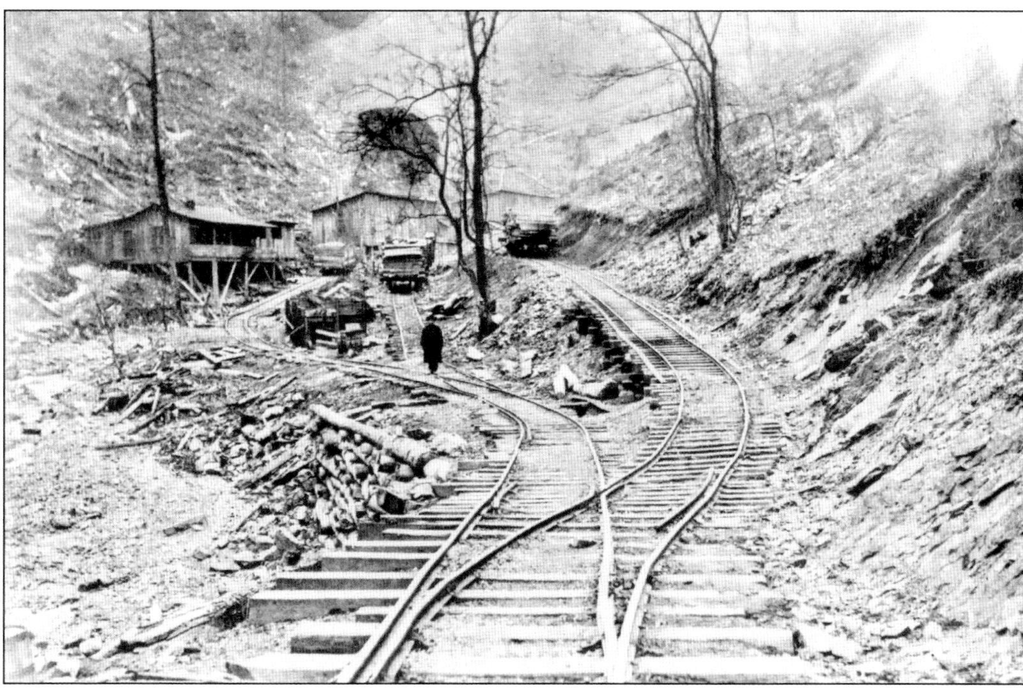

McCain Lumber Company, located near the Little Milligan community, ran its own narrow-gauge train from the logging camps to the sawmill. (Photograph courtesy of Butler Museum.)

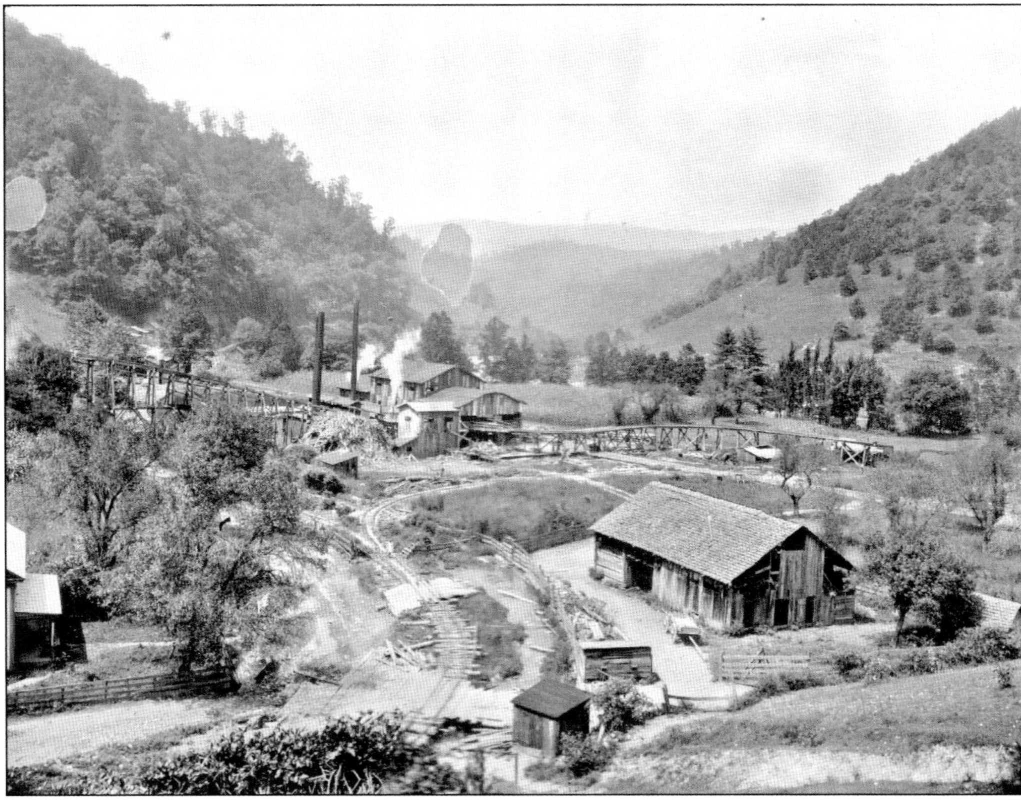

This lumber mill had a narrow-gauge track that connected the mill with its lumber camps in the region. (Photograph courtesy of Butler Museum.)

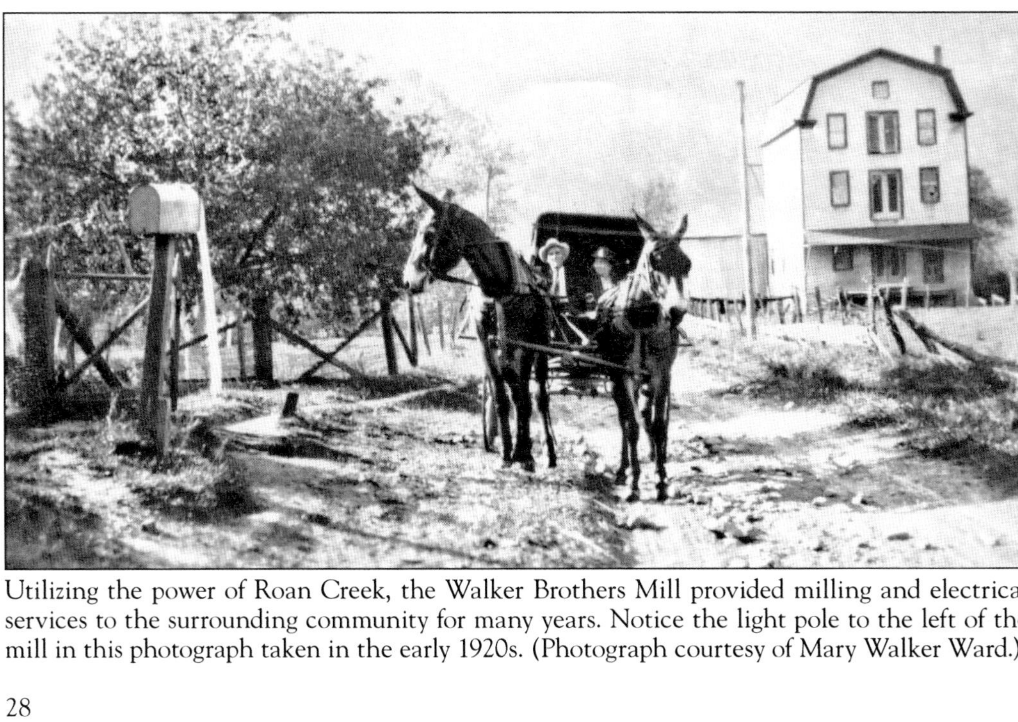

Utilizing the power of Roan Creek, the Walker Brothers Mill provided milling and electrical services to the surrounding community for many years. Notice the light pole to the left of the mill in this photograph taken in the early 1920s. (Photograph courtesy of Mary Walker Ward.)

Following World War I, the Walker Brothers Power and Light Company provided electricity for the residents of Butler from 6:00 p.m. until midnight. Pictured here is the generator that was used in conjunction with the waterwheel. (Photograph courtesy of George Walker.)

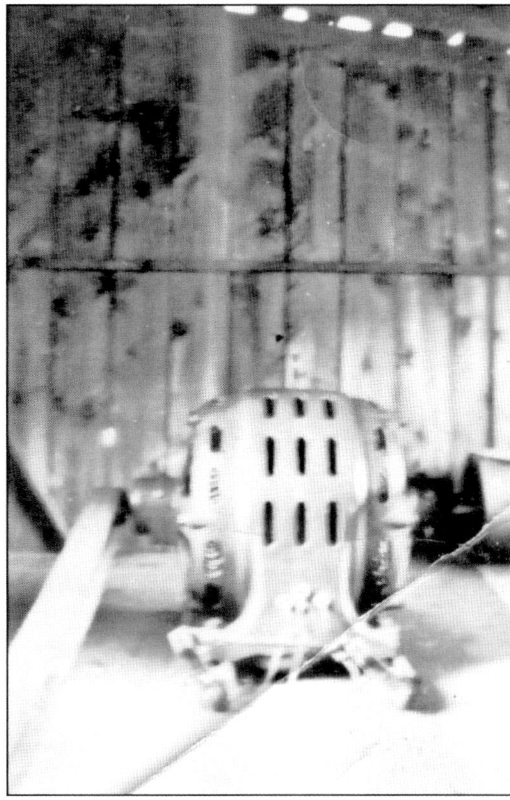

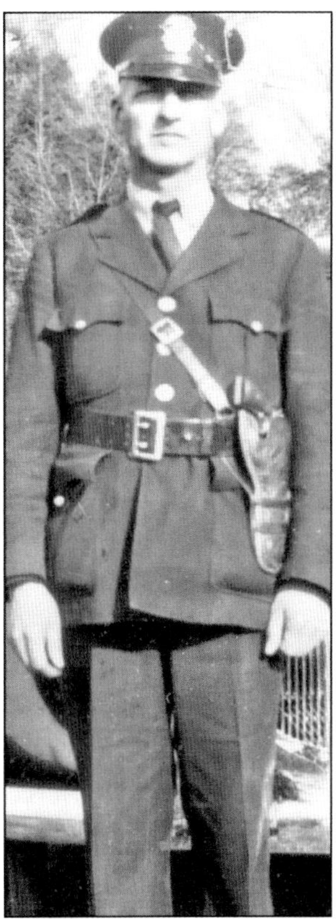

Carson Whitehead, a Spanish-American War veteran, served as a Butler policeman for many years. (Photograph courtesy of Butler Museum.)

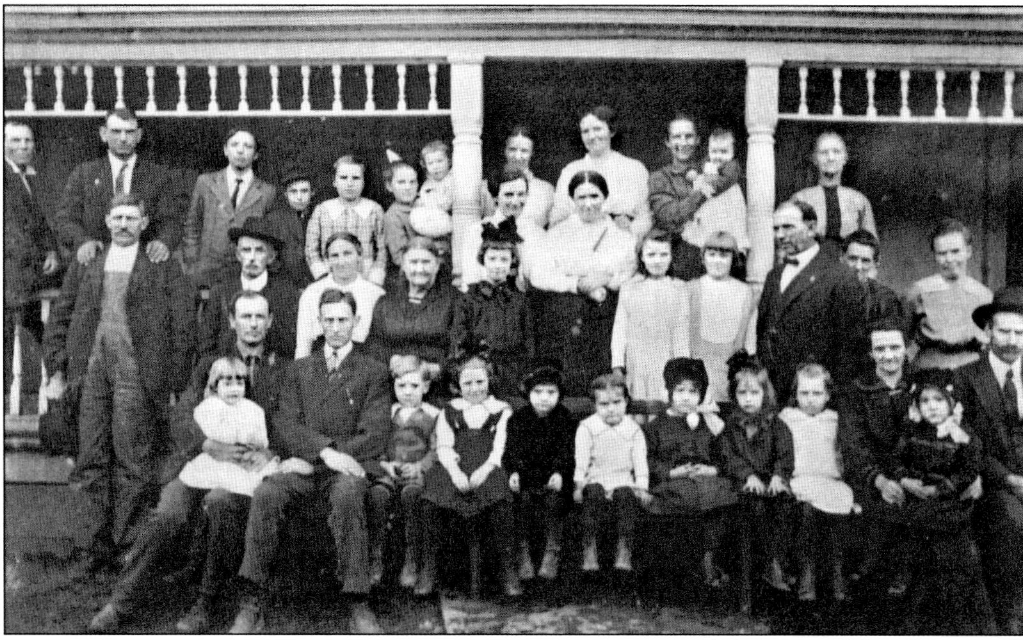

This photograph of L. L. McQueen and his family was taken in 1915. L. L. McQueen (second row, standing second from left), who was instrumental in the installation of the telephone system in Johnson County, owned the L. L. McQueen Telephone Exchange, which was sold in 1933 to Inter-Mountain Telephone Company. (Photograph courtesy of James D. Robinson.)

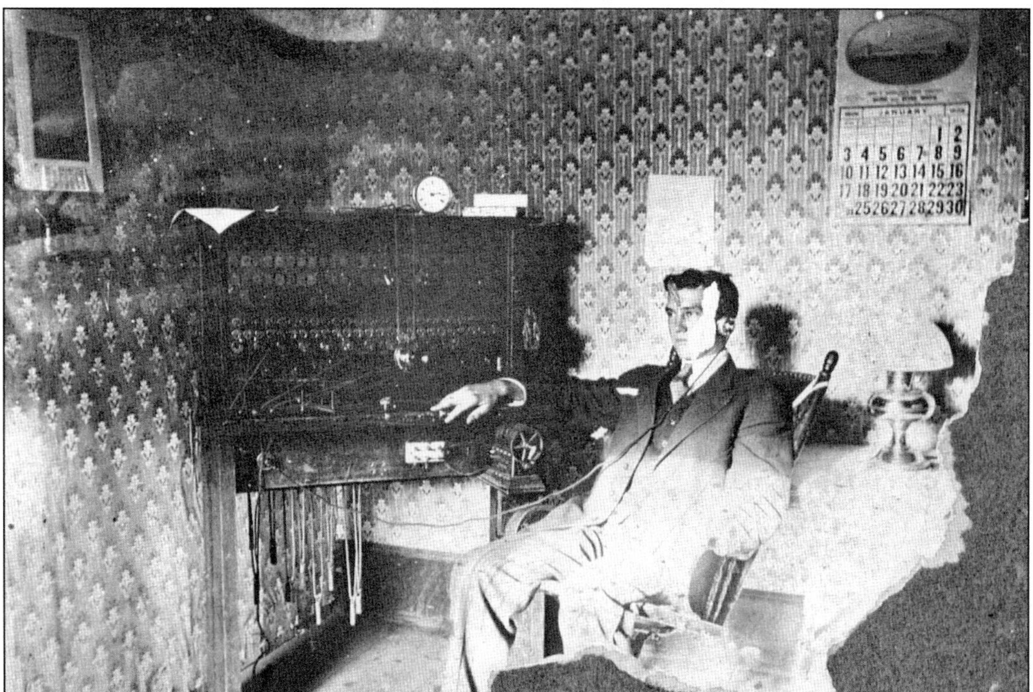

Smith McQueen, son of L. L. McQueen, mans the telephone switchboard in this picture. (Photograph courtesy of James D. Robinson.)

Dr. James D. Robinson Sr. is pictured here on his horse named "Happy" as he makes his rounds through the community. The gentleman standing near Dr. Robinson is Mr. Laws. (Photograph courtesy of James D. Robinson.)

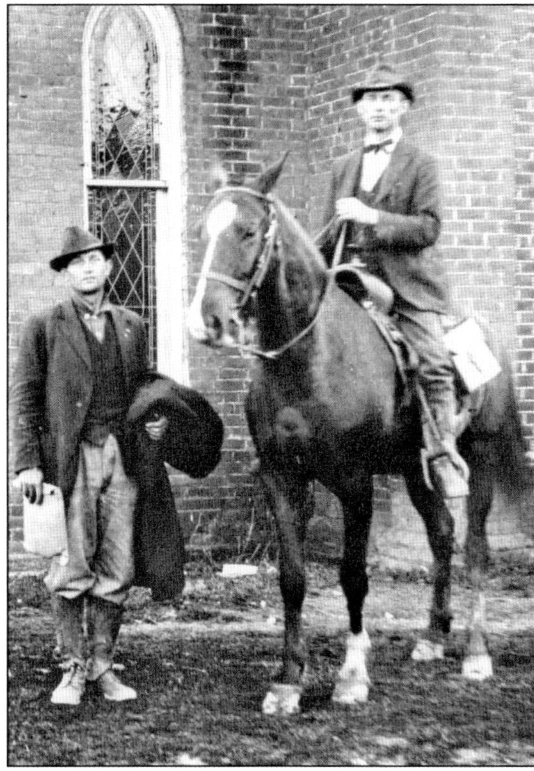

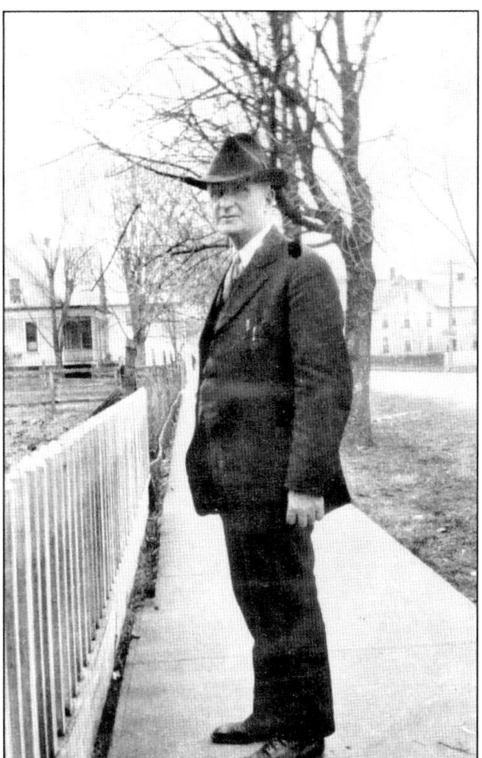

This photograph of Dr. David Swift standing on Main Street shows the Shupe Hotel at the right. During the 1920s, the hotel burned and the owners' daughter, Vera Shupe, jumped from an upstairs window, only to die later from her burns. (Photograph courtesy of Butler Museum.)

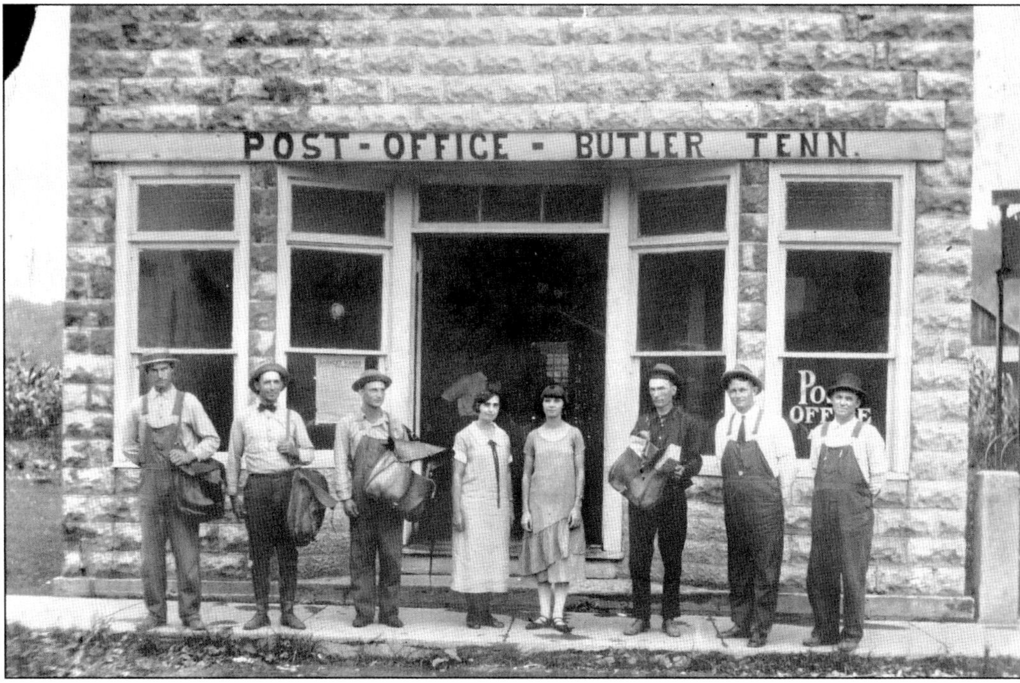

Neil Stout, Don Butler, Duff Dugger, Inez Reece, Chelsea Laws Rhudy, Clyde Lipford, postmaster Robert Laws, and Fred M. Matherly pose from left to right in front of the Butler Post Office on Main Street in 1925. (Photograph courtesy of Butler Museum.)

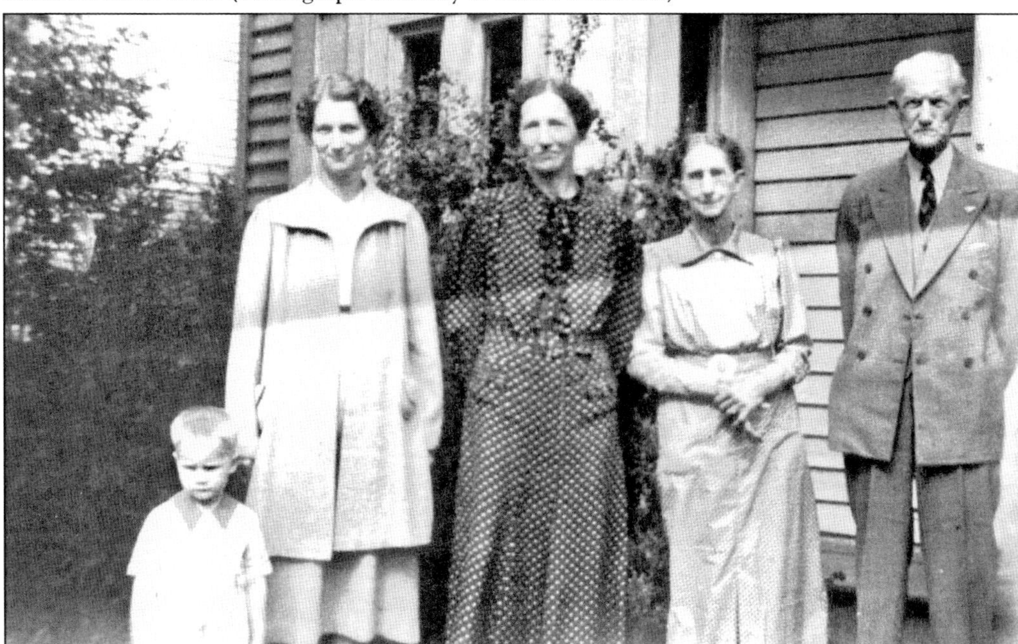

Pictured in this 1938 photograph, from left to right, are James D. Robinson, Margaret Harris Robinson, Margaret Smythe Harris, Inez Ingram Smythe, and Robert Clifford Smythe. Robert Smythe was appointed postmaster of Butler in 1902. (Photograph courtesy of James D. Robinson.)

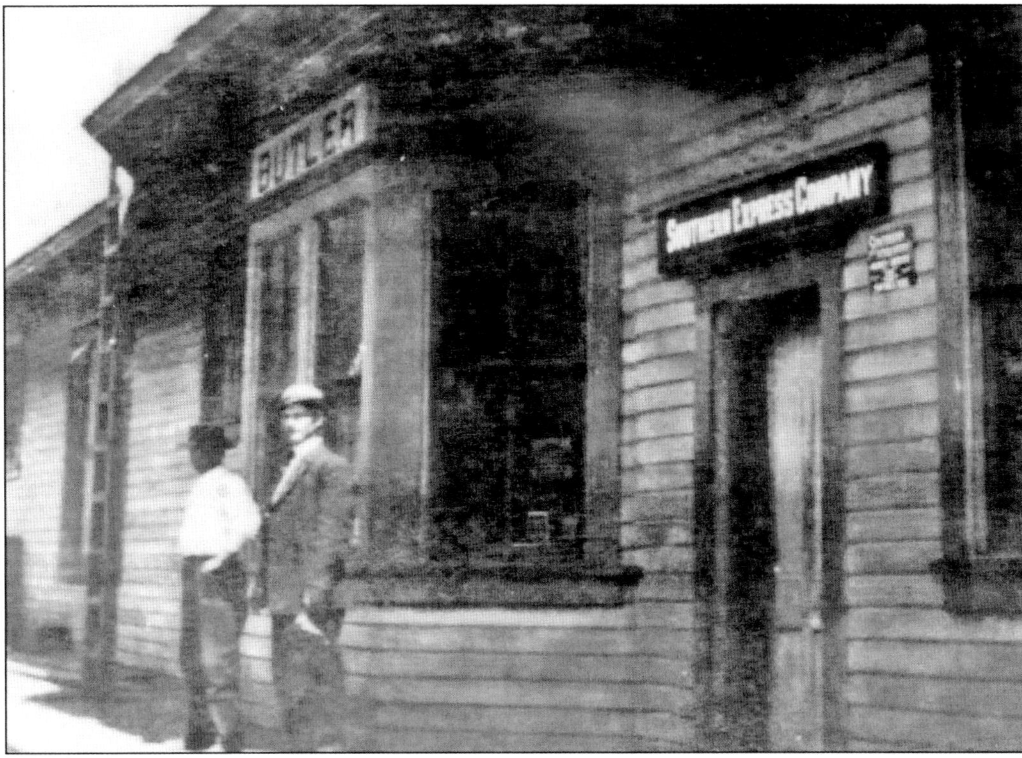

Pictured here is the office of the Southern Express Company in the Butler Depot. (Photograph courtesy of Butler Museum.)

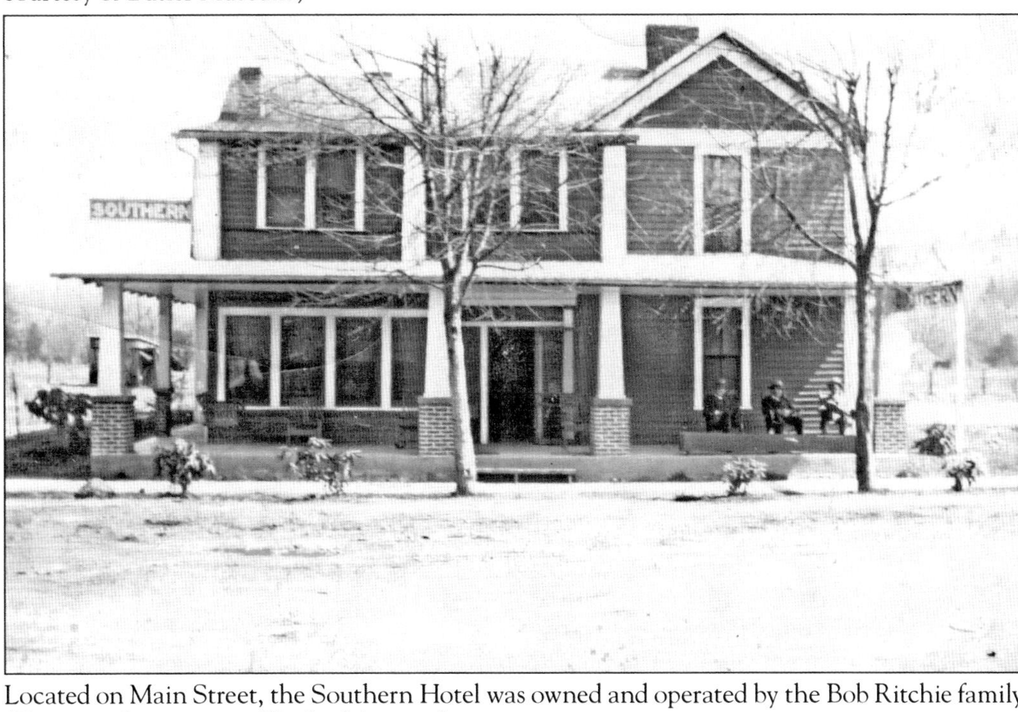

Located on Main Street, the Southern Hotel was owned and operated by the Bob Ritchie family. (Photograph courtesy of Butler Museum.)

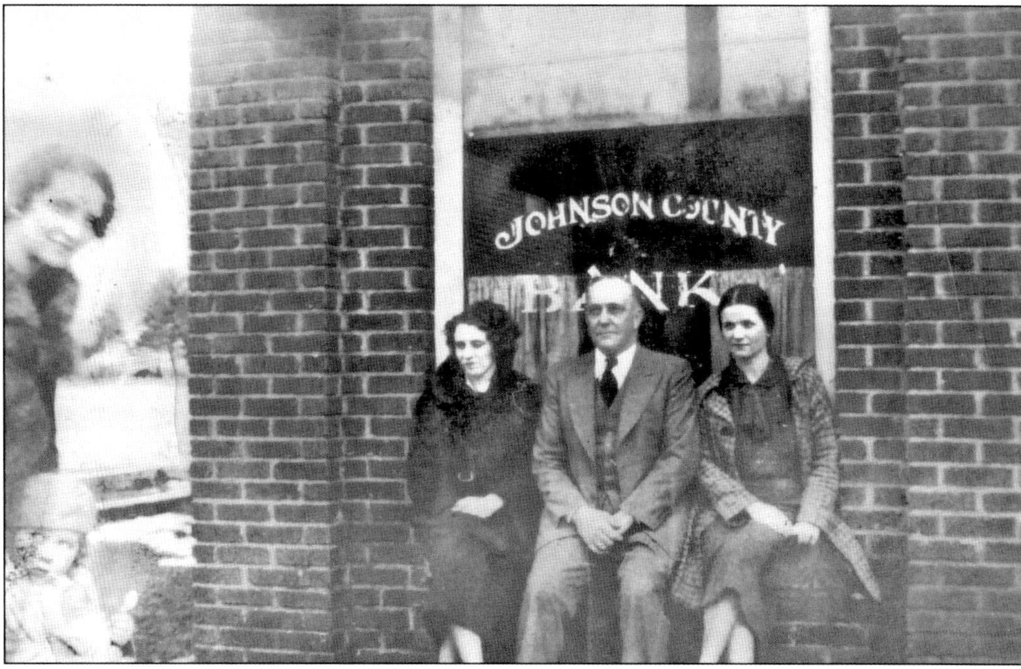

Ruth Curtis (right) sits with her co-workers at the Johnson County Bank, which served the Butler community until the stock market crash of 1929. (Photograph courtesy of Butler Museum.)

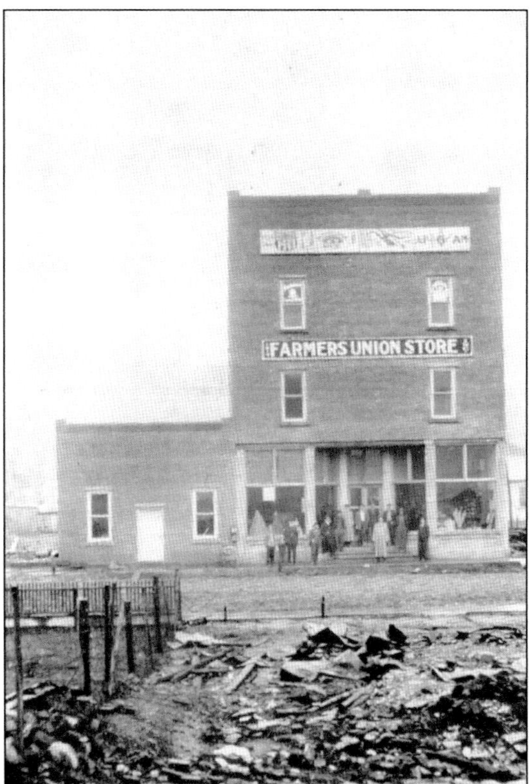

This building, which housed the Farmer's Union, functioned as several businesses throughout the years, including L. H. Goodwin's General Merchandise and Butler Furniture. The second floor of the building had a skating rink, and the third floor was utilized by the Masons, the Junior Order, and the Odd Fellows. (Photograph courtesy of Butler Museum.)

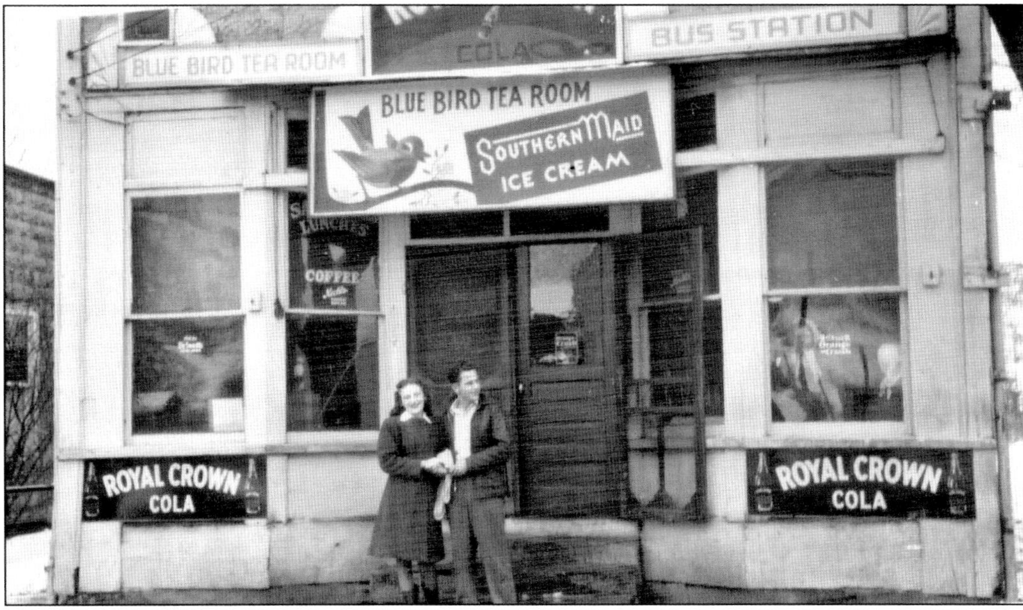

Patrons Susie Caldwell and Harold Milhorn stand in front of the Blue Bird Tea Room in this photograph taken in the 1940s. (Photograph courtesy of Butler Museum.)

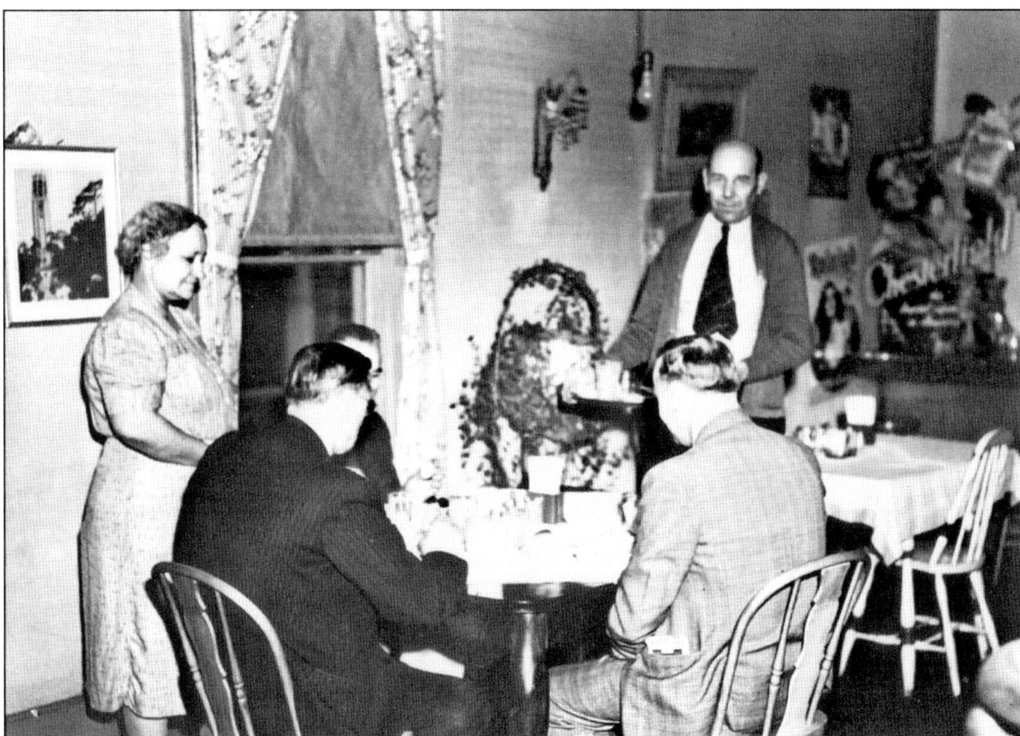

The Blue Bird Tea Room was owned and operated by Mr. and Mrs. Oliver C. Neatherly. (Photograph courtesy of Butler Museum.)

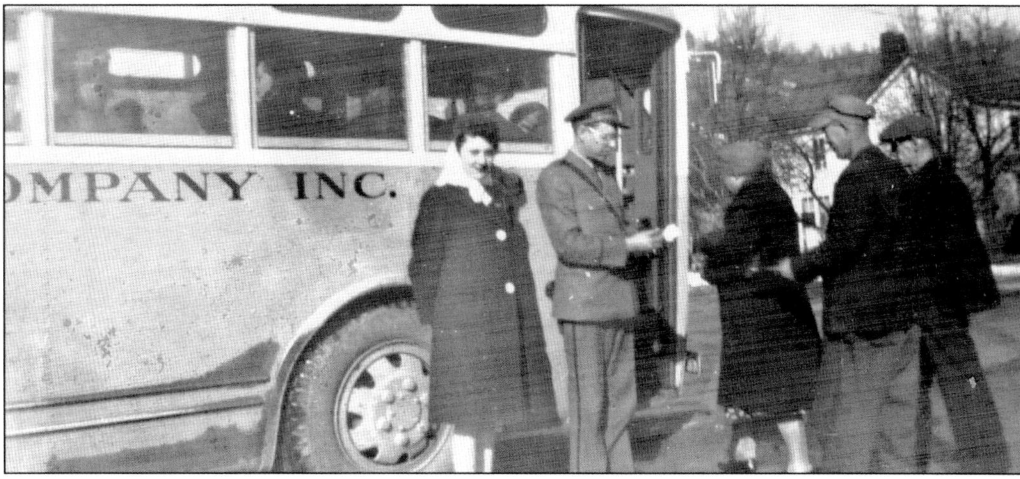

The Parkway Bus Company was owned and operated by E. O. Woody. Drew Ketron transported employees of North American Rayon Corporation and American Bemberg Corporation from Butler to Elizabethton, Tennessee. (Photograph courtesy of Butler Museum.)

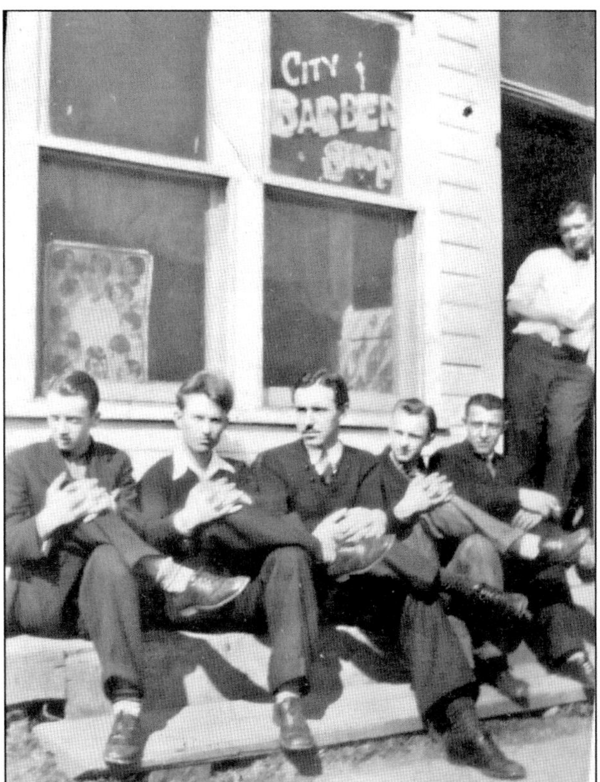

Included in this photograph of the City Barber Shop, operated by Kyle Stout, are Sproles Butler and Wayne Lipford. (Photograph courtesy of Butler Museum.)

The city hall building, originally built for Johnson County Bank, housed Butler's municipal government and police headquarters. Since 1907, the city government was made up of a mayor and a board of aldermen. Notice the three-digit telephone number on the Butler City Cab in this 1940s photograph. (Photograph courtesy of Butler Museum.)

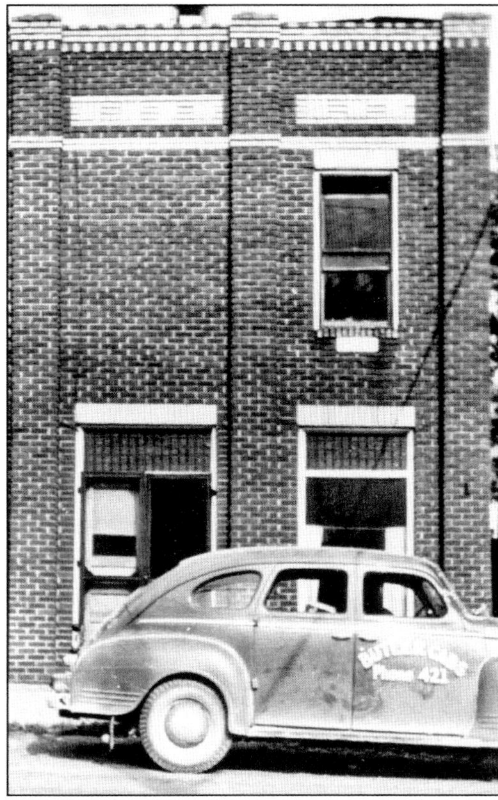

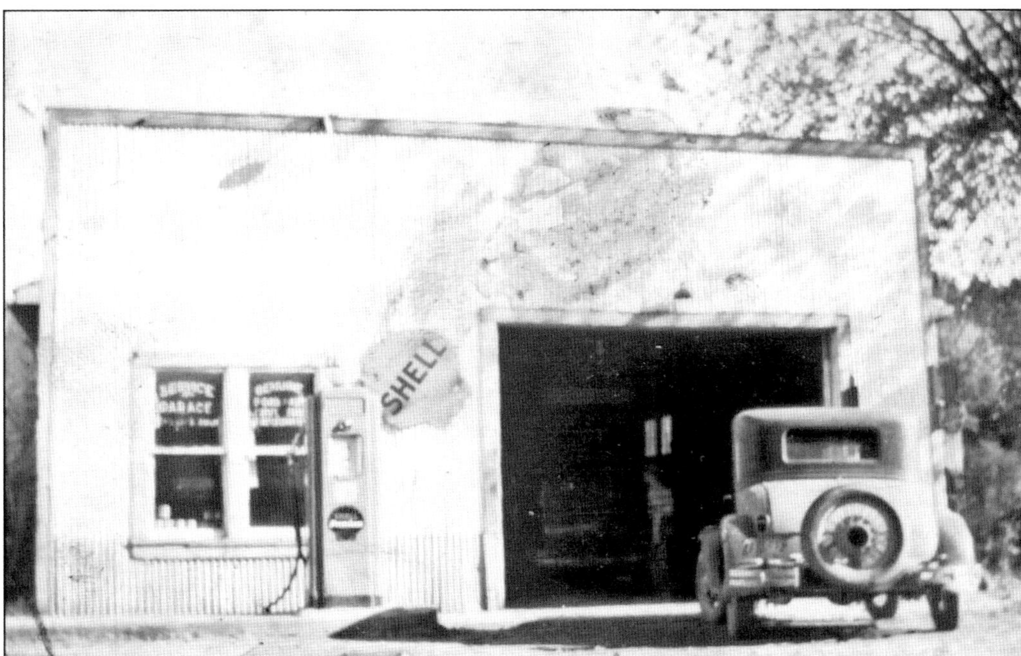

Pictured here is Ramsey's Garage, owned and operated by Hubert Ramsey and located on Main Street. (Photograph courtesy of Butler Museum.)

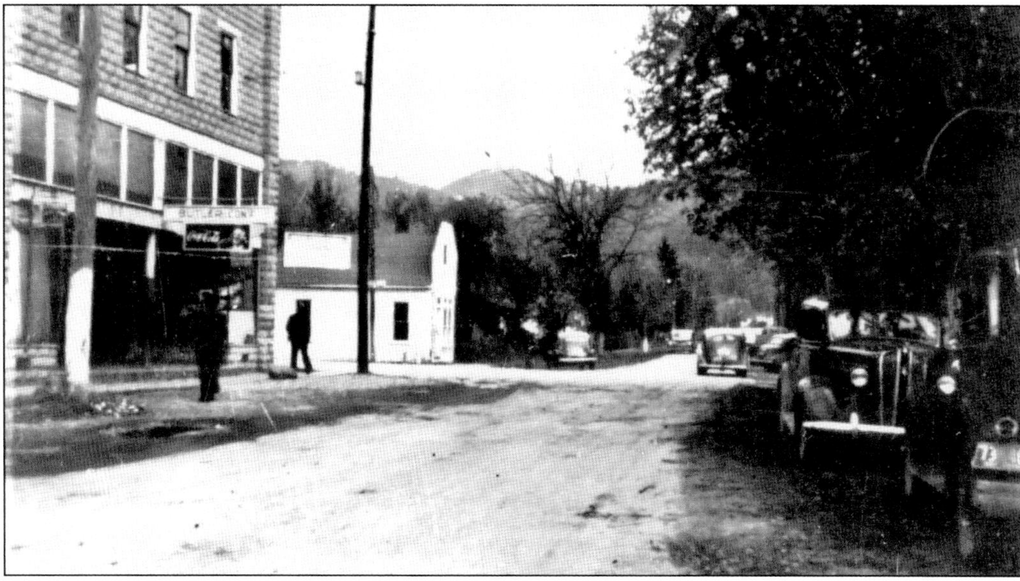

Butler Confectionary (commonly known as Butler Drugstore), operated by R. T. Smith and Ruth Curtis, started out as F. P. Curtis and Sons Store. The little white building in the background served as many businesses during the 1930s, including a watchmaker shop operated by Bosa Tribett, a restaurant, and a dentist's office operated by Dr. Horace Madron. (Photograph courtesy of Butler Museum.)

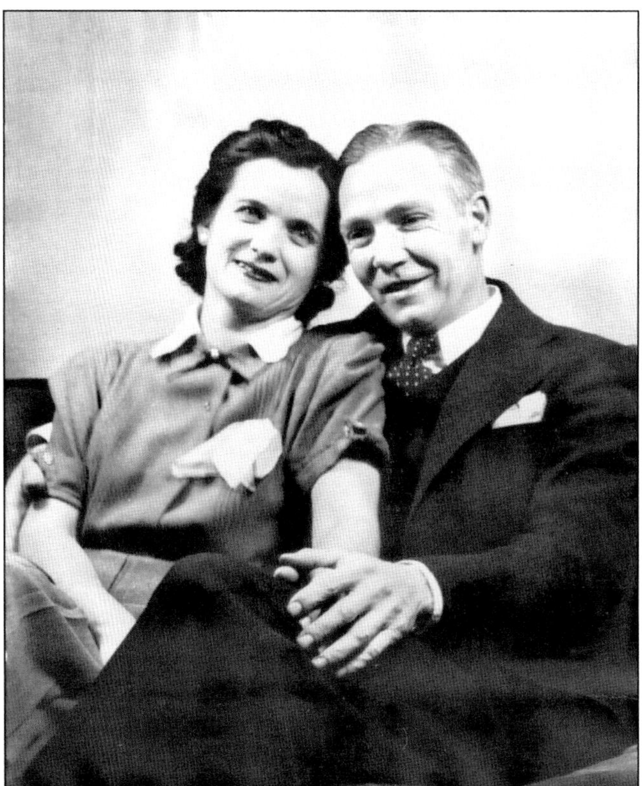

This 1940s photograph is of Finley P. and Ruth Wyatt Curtis, owners and operators of Curtis Drugstore. F. P. Curtis served as postmaster of Butler in the 1930s. (Photograph courtesy of Butler Museum.)

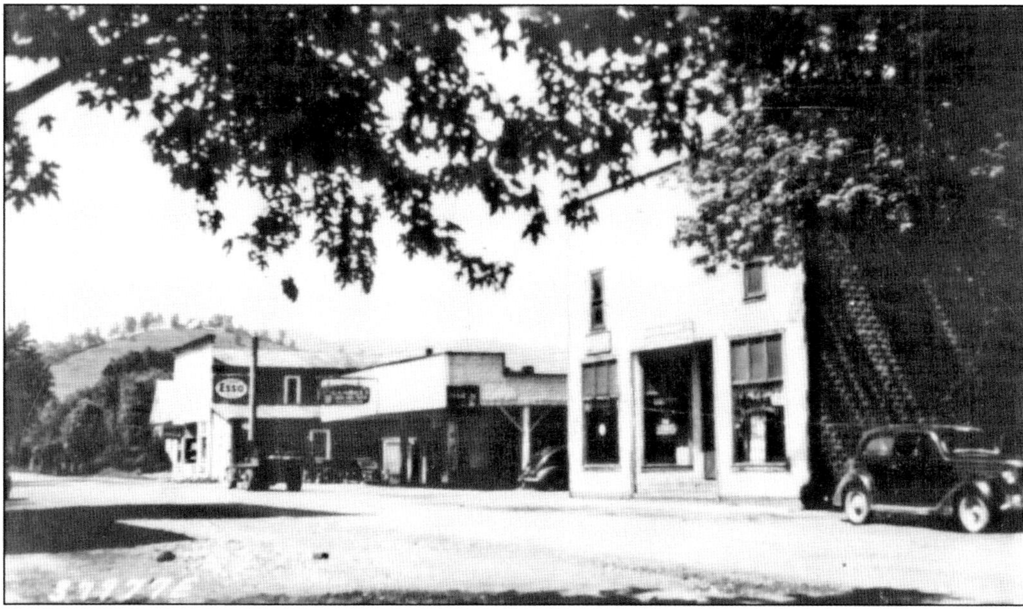

This photograph of Main Street shows Butler Hardware, owned and operated by Cliff Hampton, on the right; Bird Eye Service Station, operated by E. G. Milsap; and Rudy Ramsey's Store, which started out as L. L. Goodwin's Store, on the left. (Photograph courtesy of Butler Museum.)

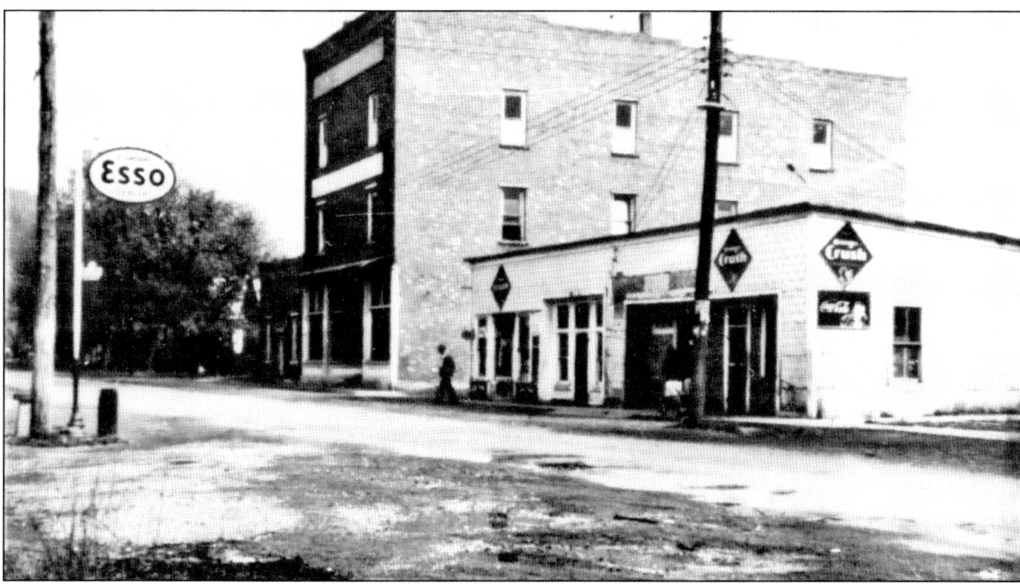

The building in the foreground at one time housed Buddy's Café, a restaurant operated by Donald Stout. Barber Lloyd Holden managed the Holden Barber Shop in the middle building, and the Farmer's Union Store was located in the building in the background. (Photograph courtesy of Butler Museum.)

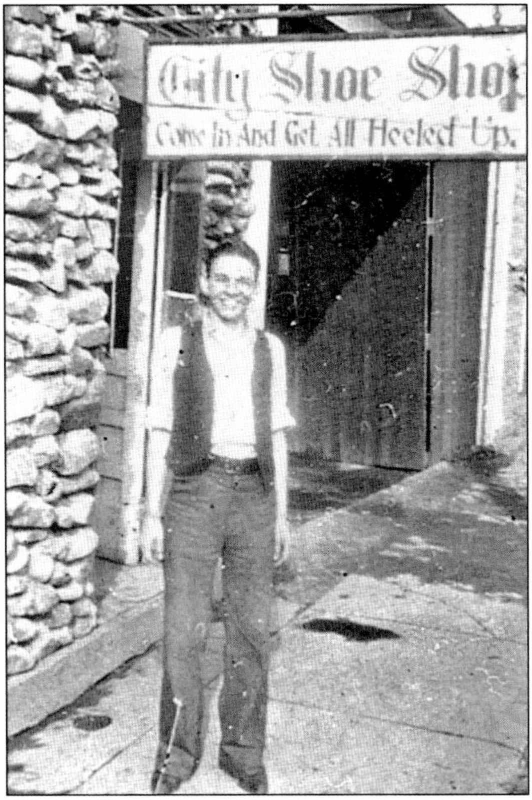

Ralph "Bud" Stout is shown here in front of the Butler Shoe Shop. (Photograph courtesy of Butler Museum.)

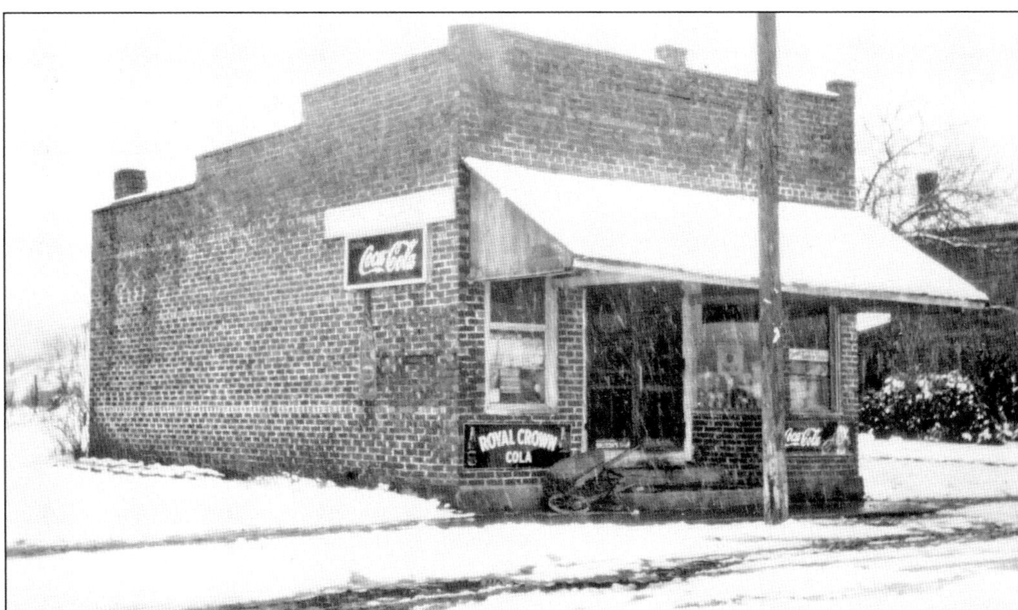

Pictured here is Alfred Stout's City Meat Market, located on Depot Street (also called Spring Street). (Photograph courtesy of Butler Museum.)

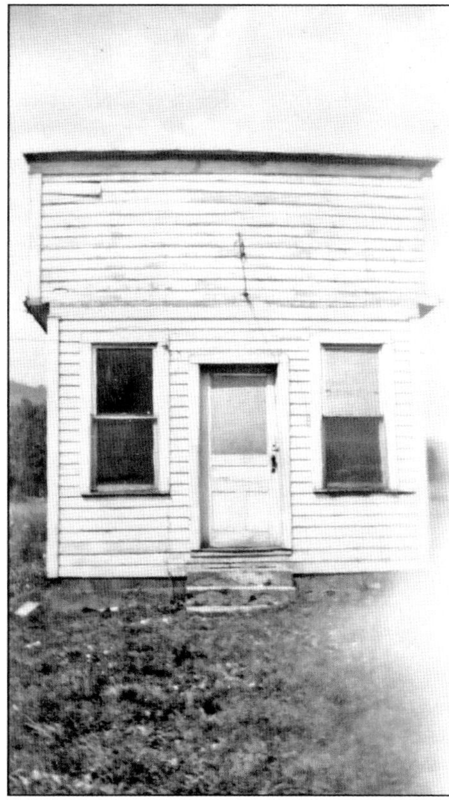

Located on College Street, the Butler city jail was locally known as the "Bug House." (Photograph courtesy of Butler Museum.)

Pictured here from left to right are Clyde Watts, Ralph Hyder, Bill Trivette, and Richard Blevins, who were members of the local Boy Scout troop in Butler. The photograph was taken in the 1940s. (Photograph courtesy of Mary Walker Ward.)

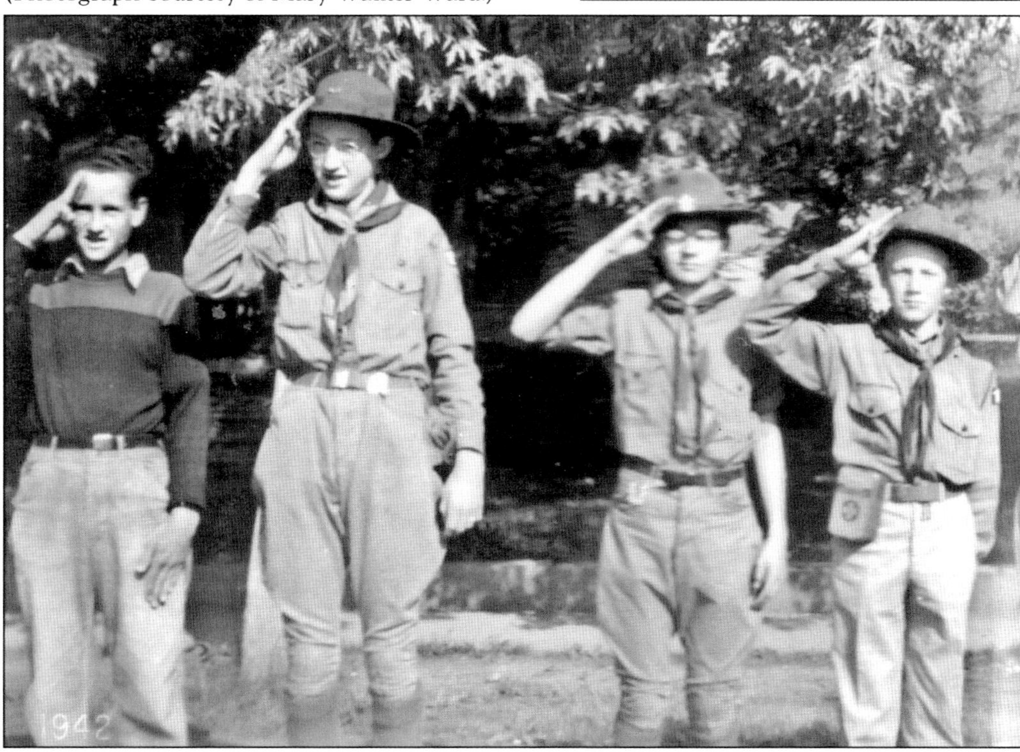

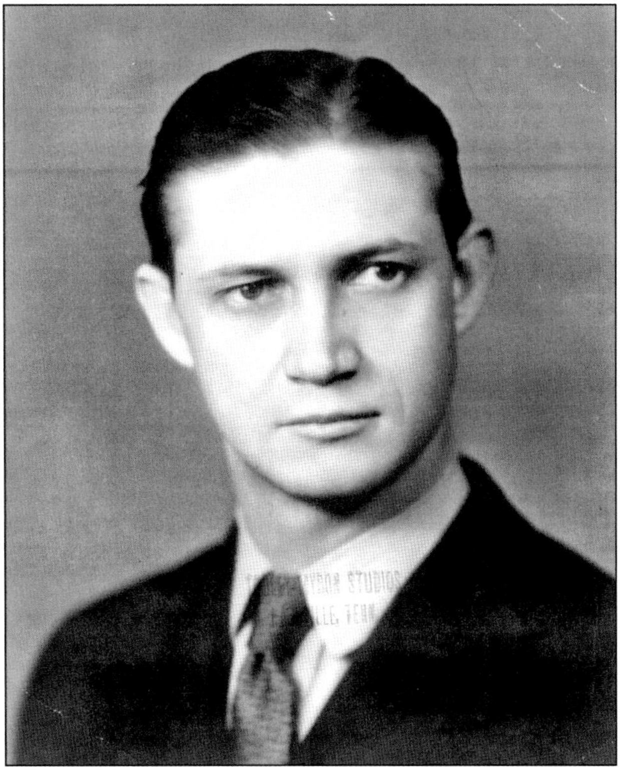

James D. Robinson Jr. was first elected to the Tennessee legislature as state representative of Johnson and Carter Counties in 1938. He served for four terms in the Tennessee General Assembly. (Photograph courtesy of James D. Robinson.)

This well-known photograph of Russell Mink walking along Main Street has been published numerous times in magazines, newspapers, and books written by local authors. The photograph was taken in 1947. (Photograph courtesy of Butler Museum.)

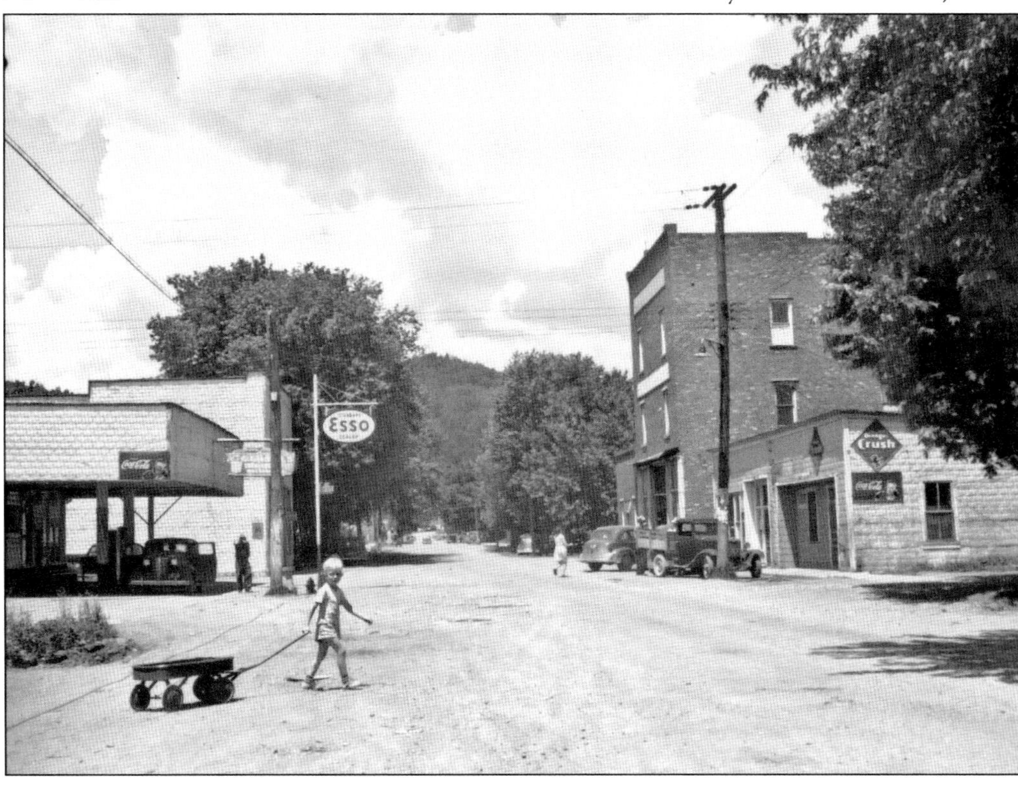

Three

Schools

In 1871, Joshua Perkins deeded the house he built for his son Frank to stockholders and trustees of Aenon (also spelled Enon) Seminary. The seminary, located near the confluence of Roan Creek and the Watauga River, established a foundation of education in the Watauga Valley that would last for more than 50 years. In 1886, under the direction of Prof. James Hamilton Smith, a new building was erected at the end of College Street that relieved the crowded conditions of the growing Aenon Seminary. During that year, the school's name was changed to Holly Spring College. The arrival of the Virginia & South Western Railroad to Butler in 1901 promoted continued growth, since students from other communities were able to travel to the school and board on campus or with local families. In 1902, the school, which was then renamed the Holly Spring Institute, became part of the Southern Baptist Board of Home Missions (a part of the division of mountain schools) and would remain so until becoming a public school in 1932. In 1906, the name was changed one final time to the Watauga Academy. Smaller schools also dotted the hillsides and communities in the Watauga Valley that met the needs of children despite their isolated location. When the Watauga Dam was built in the 1940s, the fate of these schools varied. Some, such as Little Milligan, relocated. But most closed and the students were transferred elsewhere.

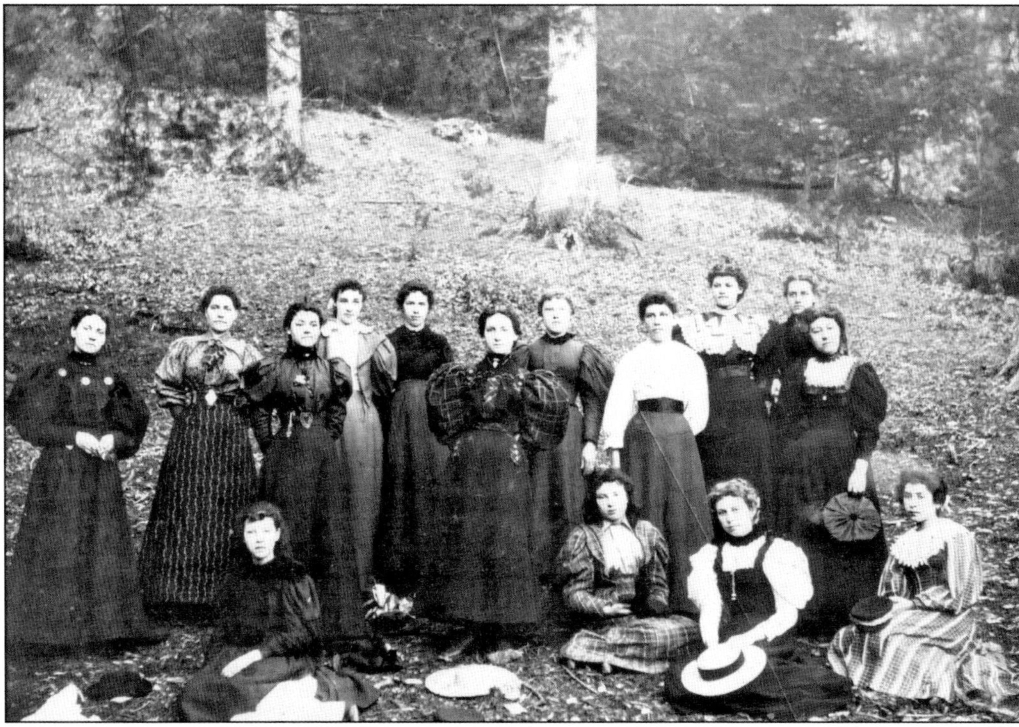

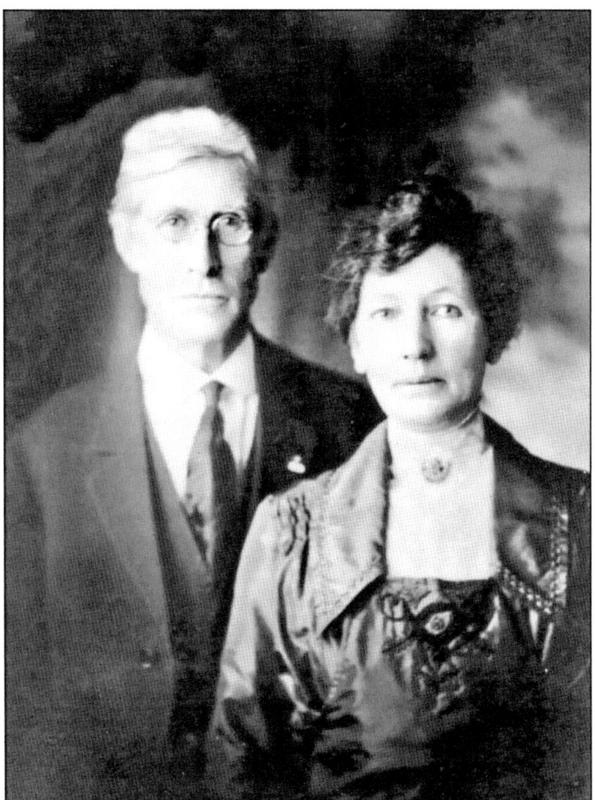

The girls of the Enon Star Literary Society pose on the grounds of the Aenon Seminary, later known as Holly Spring College. Pictured here from left to right are Lizzie Hill, Eliza Trivett, Lillie Shull, Vinnie Shull, Rebekah McQueen, Sallie Smith, Mildred Hill, Mollie Goodwin, Lettie Lineback, Willie Brummit, Crete Vaugh, Virginia (Nettie) Shull, Cora Hagaman, Bruce Slemp, and Lockie Lineback. The photograph was taken in 1885. (Photograph courtesy of Butler Museum.)

Prof. James H. Smith, graduate of Milligan College, began a fruitful career teaching in Butler in 1882. For more than 20 years, Professor Smith worked hard to administer and develop the educational tenure in the Watauga Valley. James married Mollie Shull, daughter of David and Martha Lewis Shull, in 1889. (Photograph courtesy of Butler Museum.)

This grade report of L. V. Peters, dated February 27, 1887, was signed by Prof. James Smith. (Document courtesy of Butler Museum.)

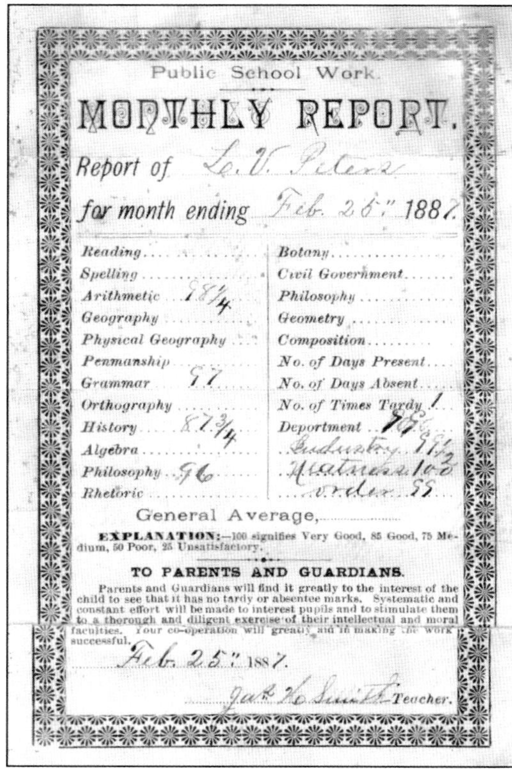

Public School Work.

MONTHLY REPORT.

Report of L. V. Peters

for month ending Feb. 25" 1887

Reading		Botany	
Spelling		Civil Government	
Arithmetic	88¼	Philosophy	
Geography		Geometry	
Physical Geography		Composition	
Penmanship		No. of Days Present	
Grammar	97	No. of Days Absent	
Orthography		No. of Times Tardy	1
History	87¾	Deportment	98
Algebra		Industry	99½
Philosophy	96	Neatness	100
Rhetoric		Order	99

General Average,

EXPLANATION:—100 signifies Very Good, 85 Good, 75 Medium, 50 Poor, 25 Unsatisfactory.

TO PARENTS AND GUARDIANS.

Parents and Guardians will find it greatly to the interest of the child to see that it has no tardy or absentee marks. Systematic and constant effort will be made to interest pupils and to stimulate them to a thorough and diligent exercise of their intellectual and moral faculties. Your co-operation will greatly aid in making the work successful.

Feb. 25" 1887.

Jas. H. Smith Teacher.

Located on Watauga River, Aenon Seminary served students of all ages from 1871 to 1886. (Photograph courtesy of Butler Museum.)

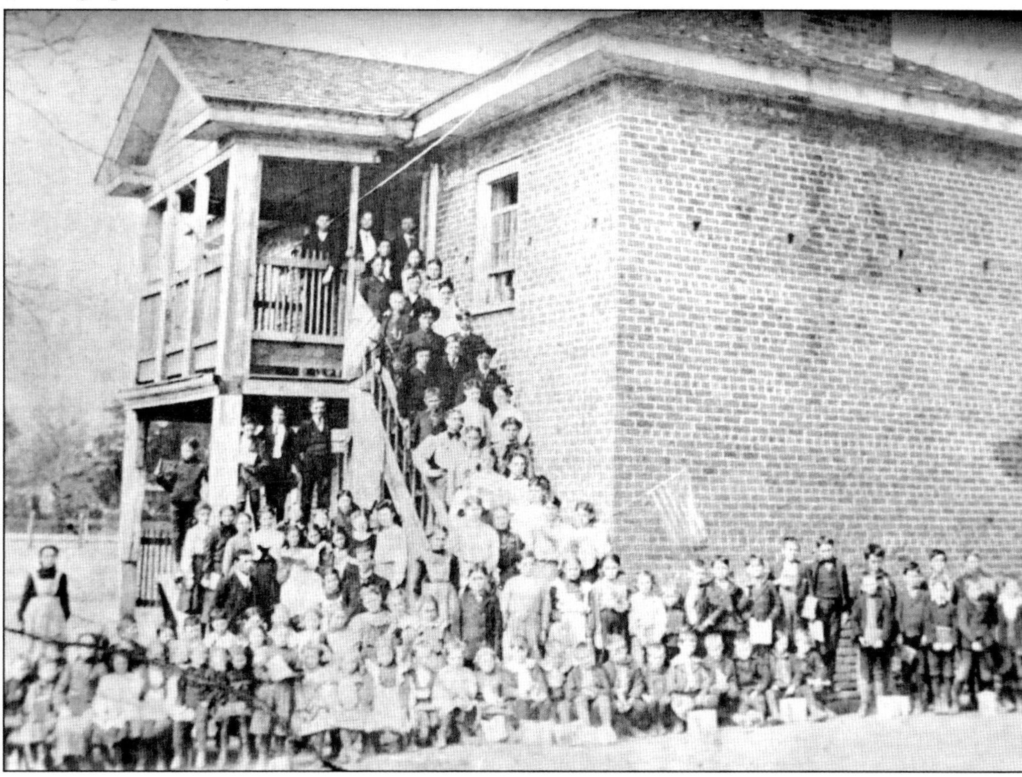

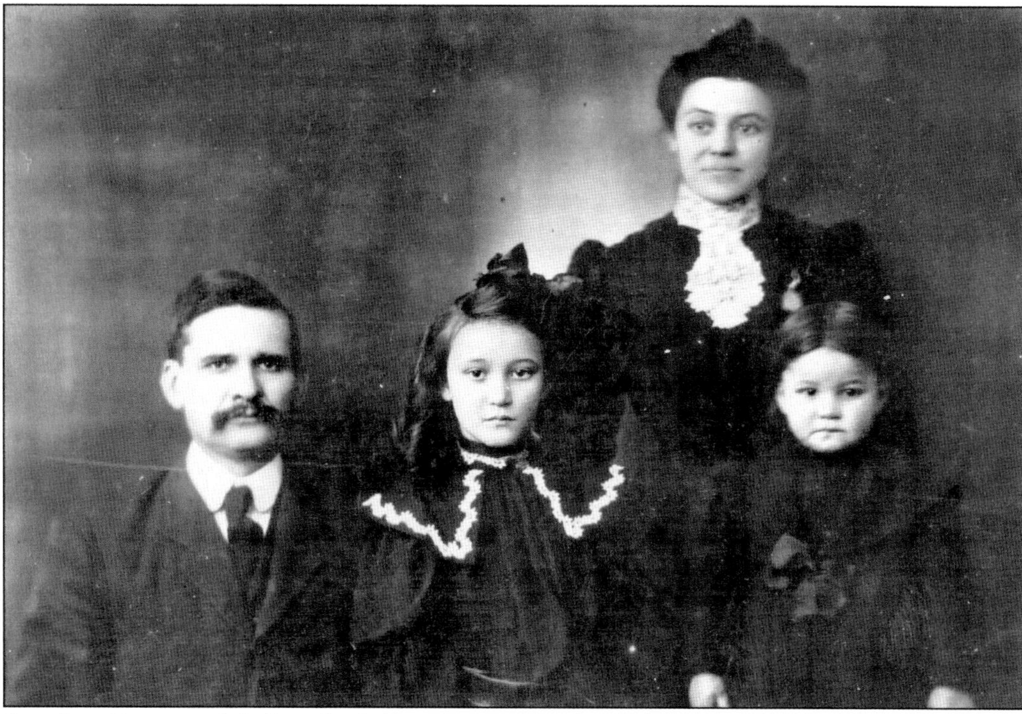

Dauphin Disco and Lillie Shull Dougherty are pictured here with their children, Clara Bartlett (left) and Annie Lewis (right). Dauphin, a professor at Holly Spring College, founded Appalachian State University in Boone, North Carolina, with his brother B. B. Dougherty. (Photograph courtesy of Butler Museum.)

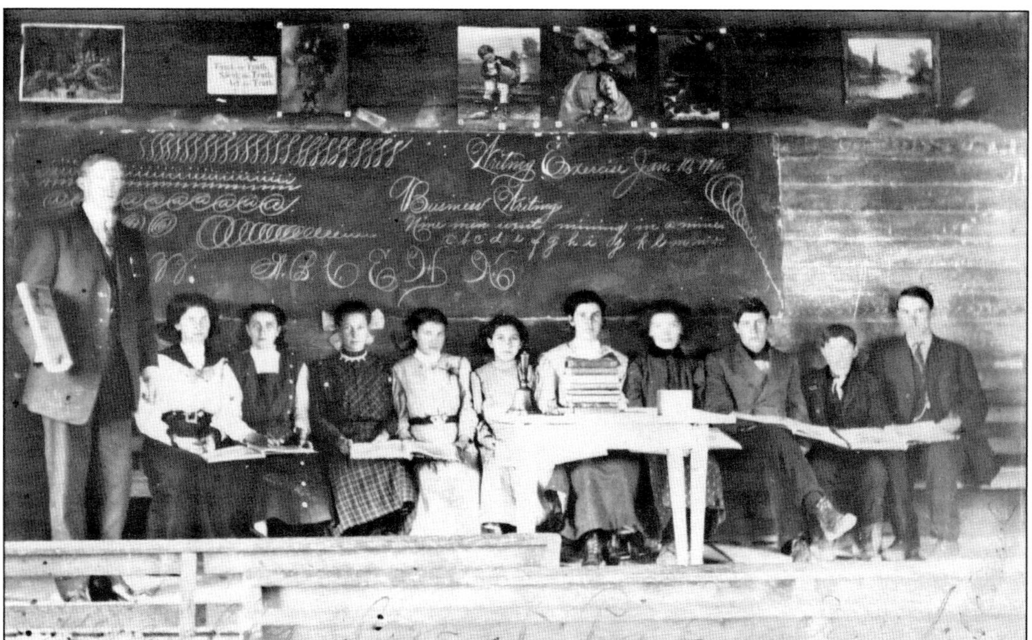

This photograph of a business-writing class at the Watauga Academy was taken in 1911. (Photograph courtesy of Butler Museum.)

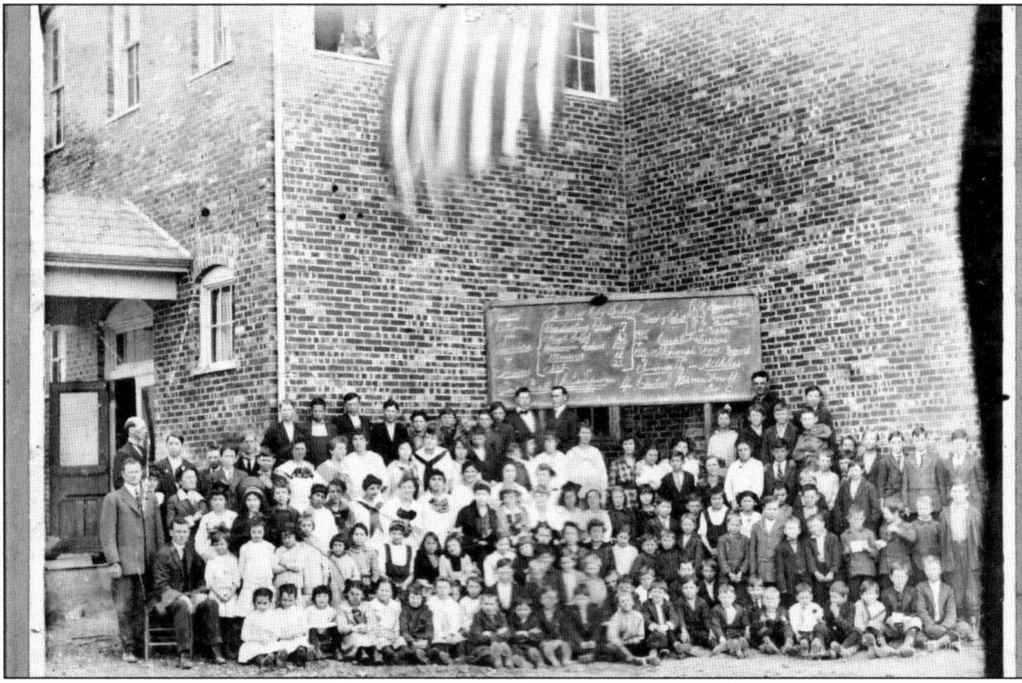

Students and faculty of Butler City School and Watauga Academy are pictured here in 1915. The chalkboard situated in the back has information about the 1915–1916 school year, including enrollment in the school (broken down by grade levels), the board of education, and athletic records. (Photograph courtesy of Butler Museum.)

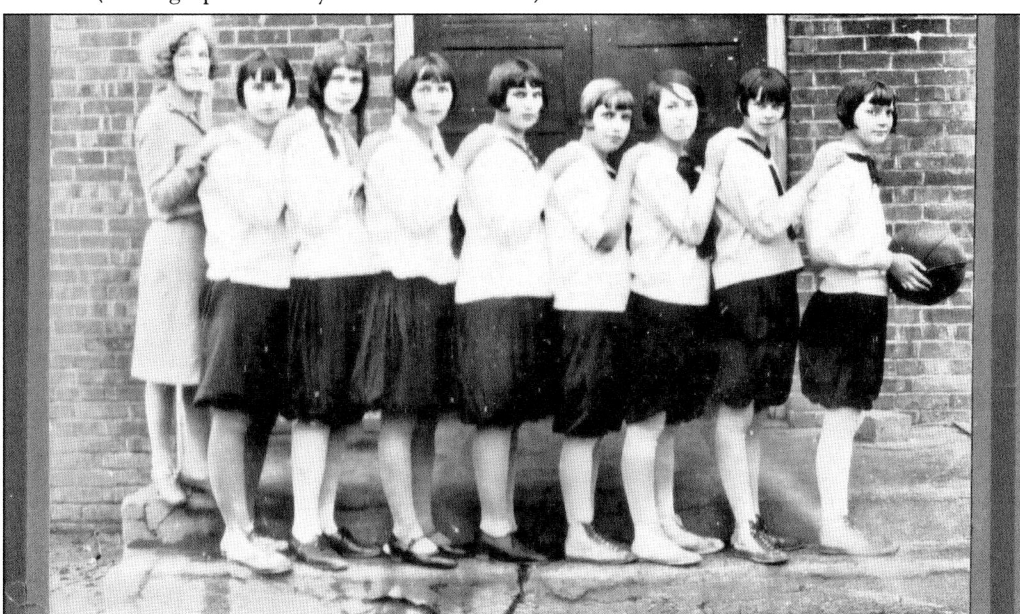

This photograph of the Watauga Academy girls' basketball team was taken *c.* 1920. The girls are standing on the front steps of Watauga Academy. The elaborate front porch that was present when the building was torn down in the 1940s is absent in this image; it was built in 1924. (Photograph courtesy of Butler Museum.)

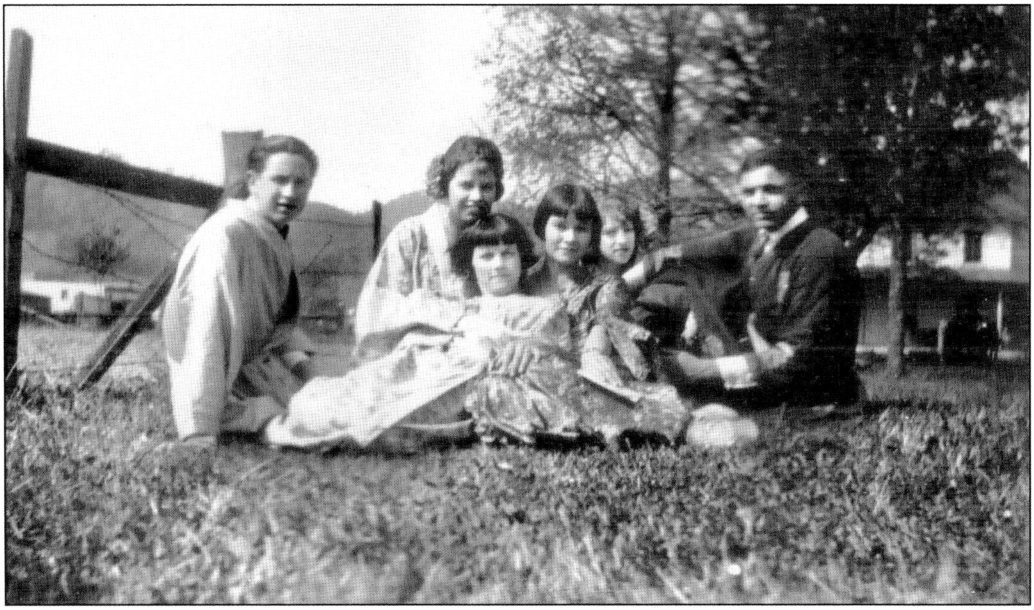

On May 6, 1925, the senior class of Watauga Academy presented A *Japanese Tea* during commencement week. The cast of characters pictured here from left to right are Sam Perry, Ethel Dugger, Mary Goodwin, Ruth Goodwin, Aileen King, and unidentified. (Photograph courtesy of Butler Museum.)

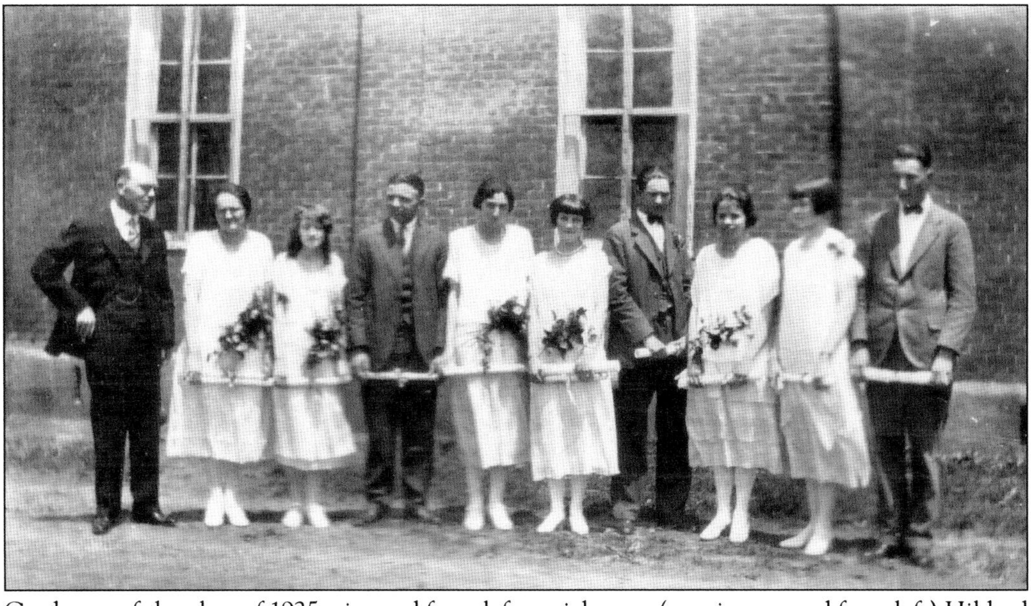

Graduates of the class of 1925, pictured from left to right, are (starting second from left) Hildred Wagner, Rheta Williams, Sam Perry, Aileen King, Mary Goodwin, Earl Neatherly, Ethel Dugger, Ruth Goodwin, and Robert Grindstaff. Prof. C. C. Perry is at the left. (Photograph courtesy of Butler Museum.)

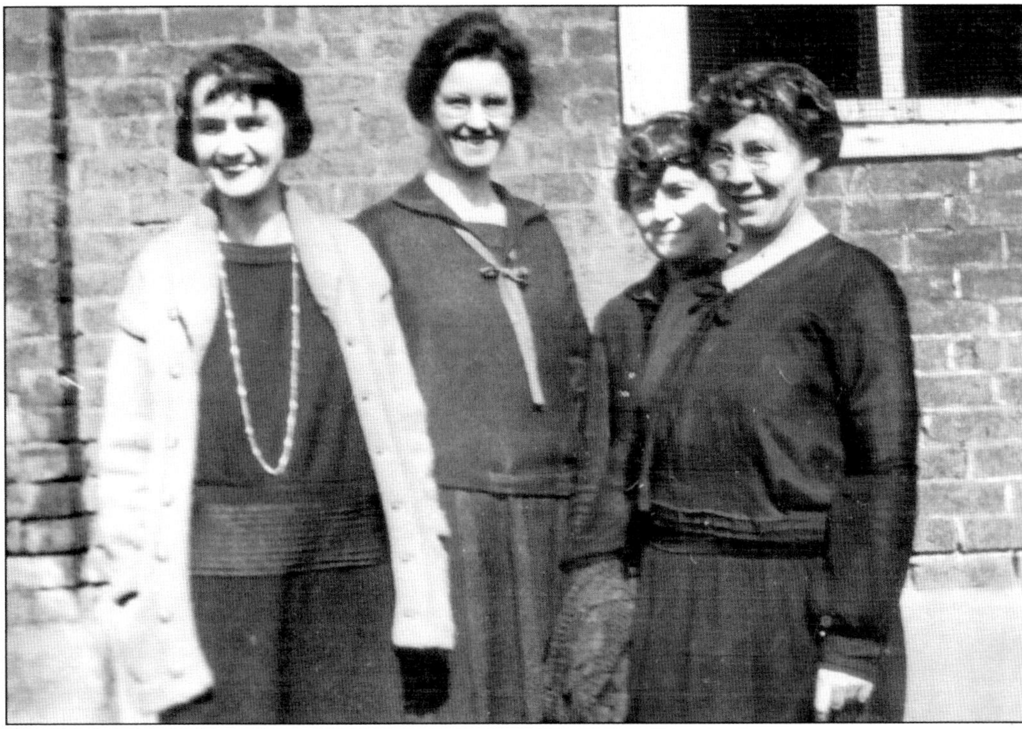

Pictured here from left to right are Elizabeth Watkins, Loretta Stout, Anna J. Merryman, and Betty Thornton, teachers at Watauga Academy in 1925. (Photograph courtesy of Butler Museum.)

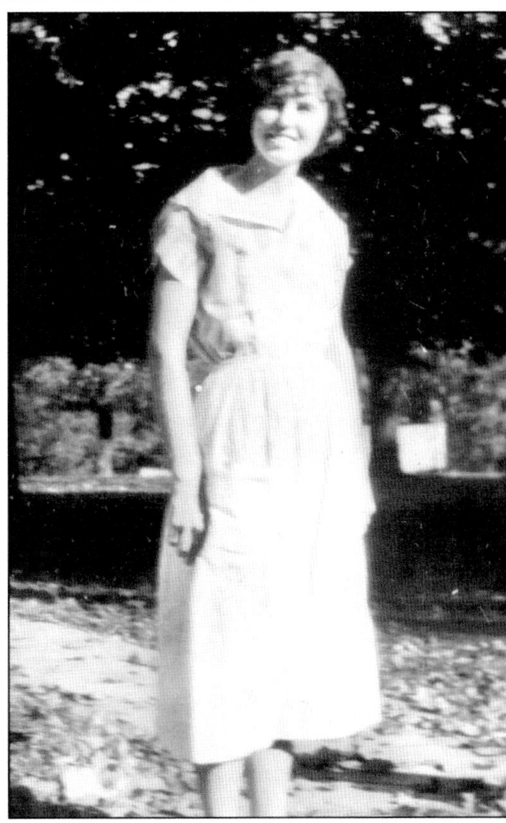

Lucia Burnett served as a home economics teacher at Watauga Academy in 1925. (Photograph courtesy of Butler Museum.)

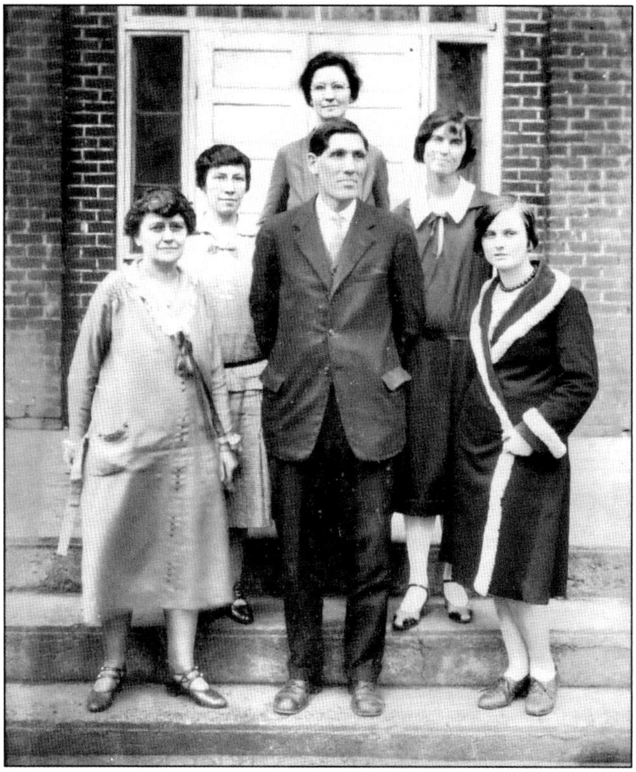

The faculty of Watauga Academy around 1925 included principal J. L. Underwood, Mrs. J. L. Underwood, Miss Loretta Stout, Miss Anna J. Merryman, Miss Bettie Thornton, and Mrs. J. A. Slemp. (Photograph courtesy of Butler Museum.)

This photograph of the students and faculty of Watauga Academy was taken during the early 1920s. (Photograph courtesy of Butler Museum.)

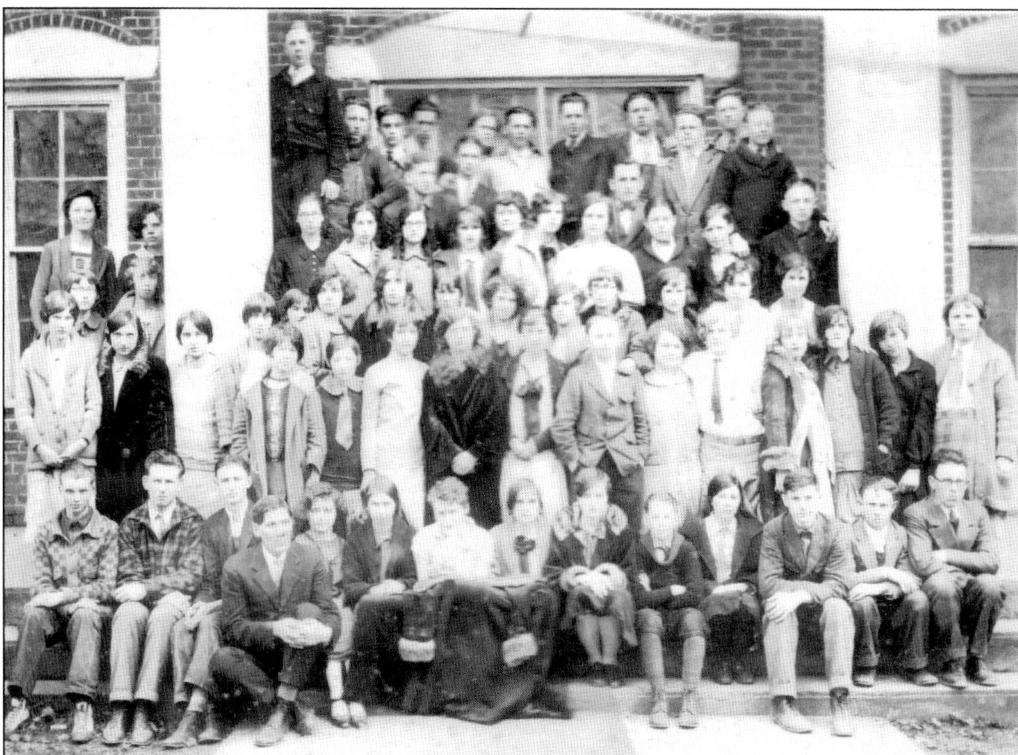

C. A. Todd served simultaneously as principal of Watauga Academy and as pastor of Butler Baptist Church from 1927 to 1935. (Photograph courtesy of Butler Museum.)

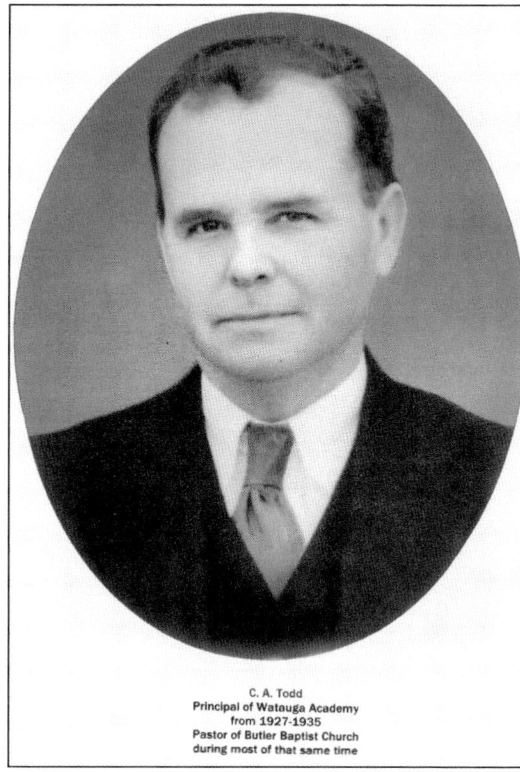

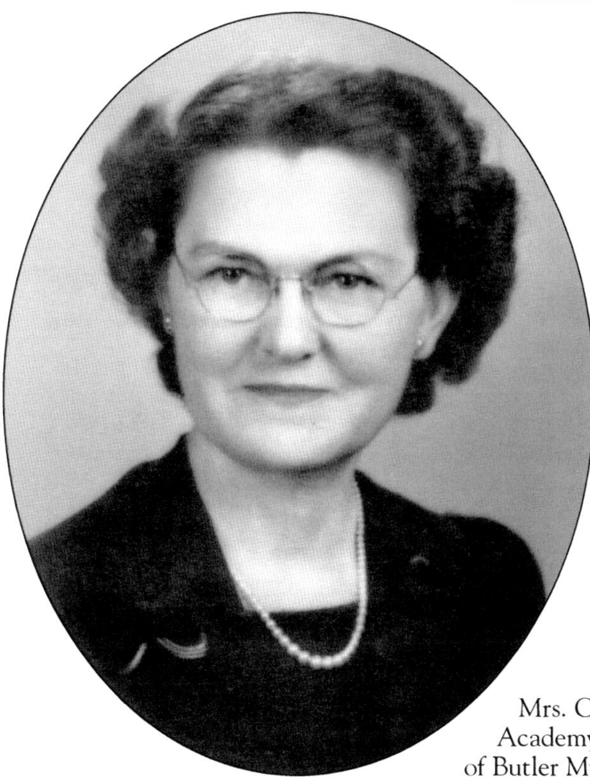

Mrs. C. A. Todd served as librarian of Watauga Academy from 1927 to 1935. (Photograph courtesy of Butler Museum.)

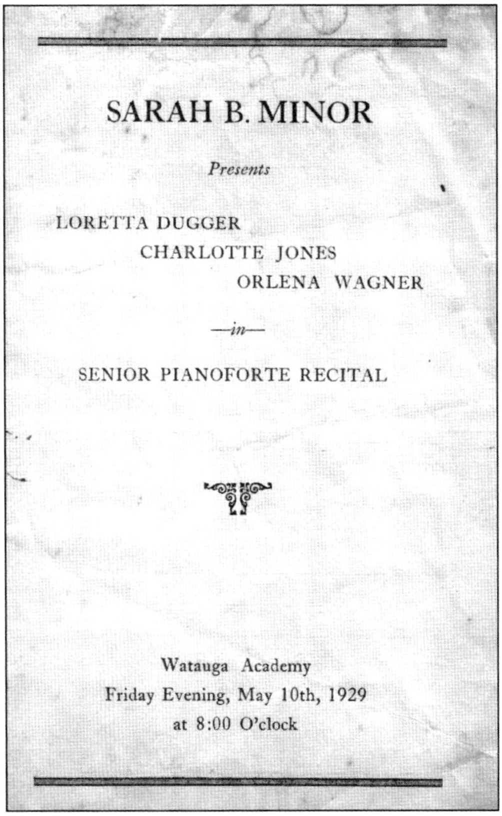

SARAH B. MINOR

Presents

LORETTA DUGGER
CHARLOTTE JONES
ORLENA WAGNER

—*in*—

SENIOR PIANOFORTE RECITAL

Watauga Academy
Friday Evening, May 10th, 1929
at 8:00 O'clock

The recital of Loretta Dugger, Charlotte Jones, and Orlena Wagner in 1929 included selections from Beethoven, Torjussen, Schuman, and Leschetizky. (Document courtesy of Butler Museum.)

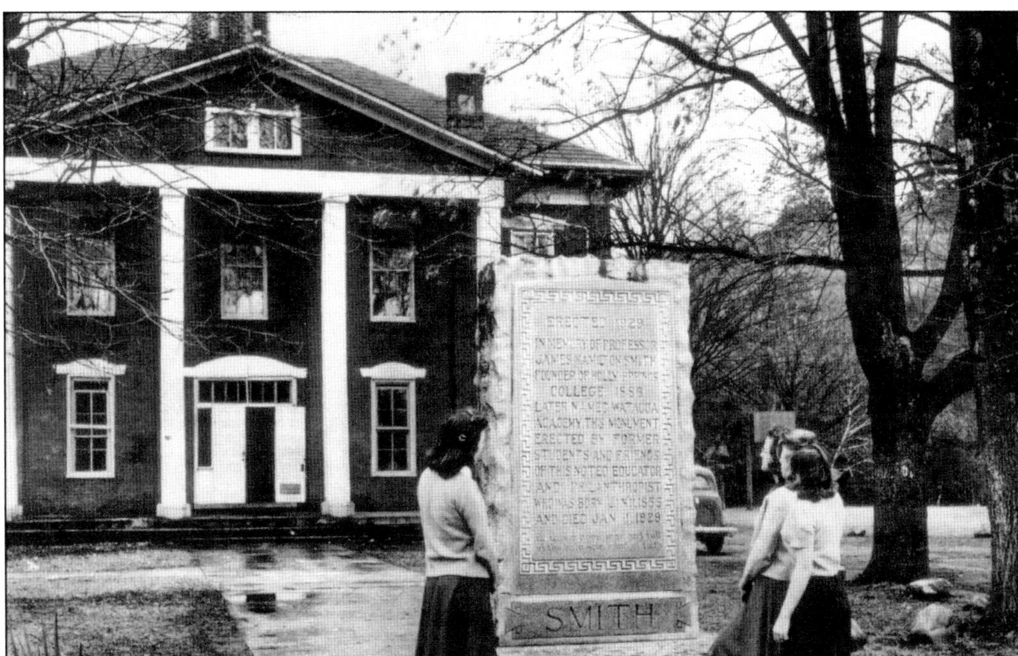

The Smith Monument was erected and dedicated on the campus of Watauga Academy in 1929. It honored Prof. James Smith. (Photograph courtesy of Butler Museum.)

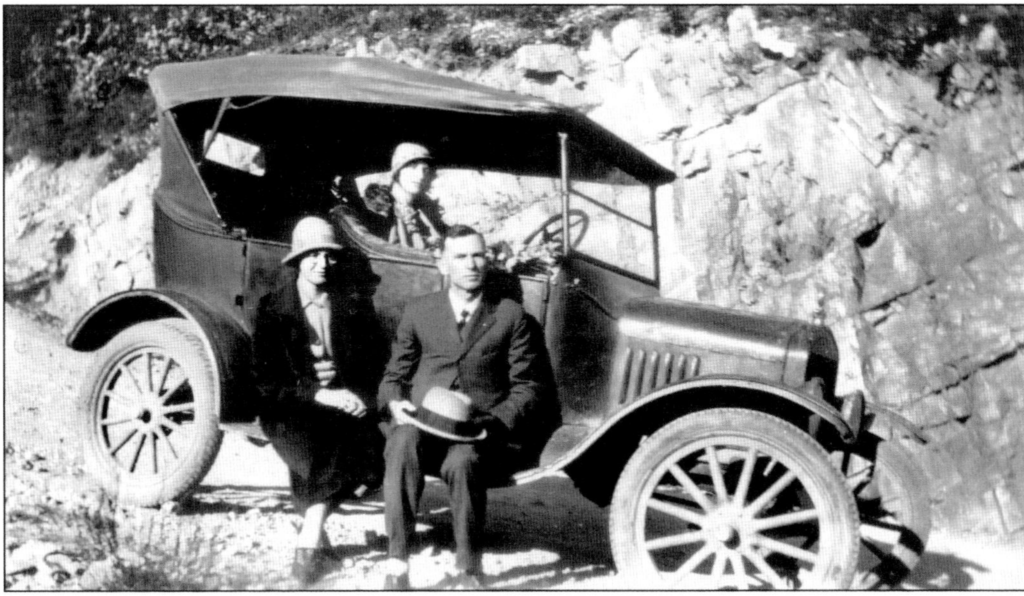

Robert and Bess DeVault pose on the running board of their first car. Robert M. DeVault Sr. served as the principal of Watauga Academy and as pastor of Butler Baptist Church. (Photograph courtesy of Butler Museum.)

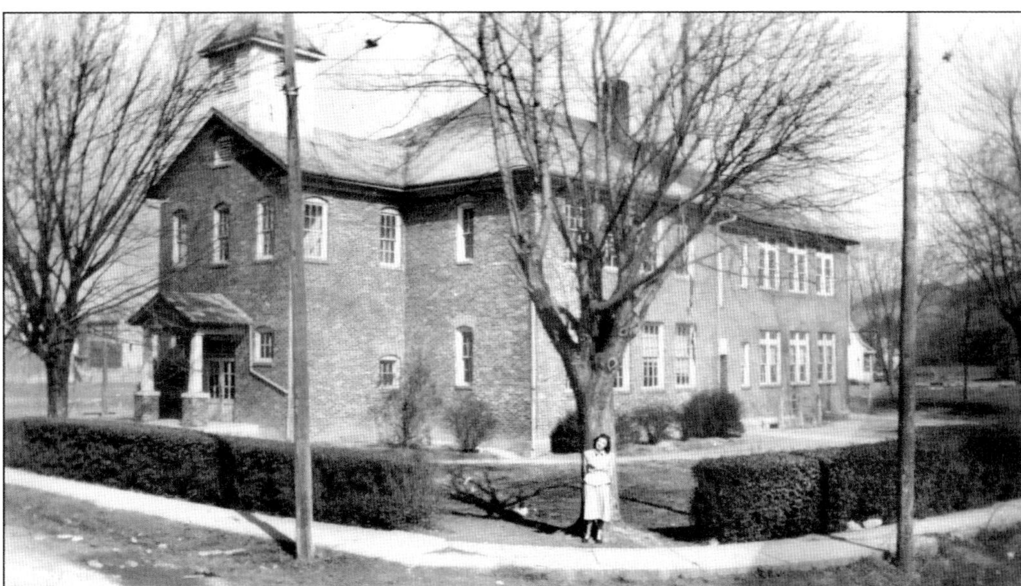

Lucille McQueen stands in front of Butler City School. Located on the corner of School and College Streets, Butler City School educated elementary students from 1908 until it was demolished in 1948. (Photograph courtesy of Butler Museum.)

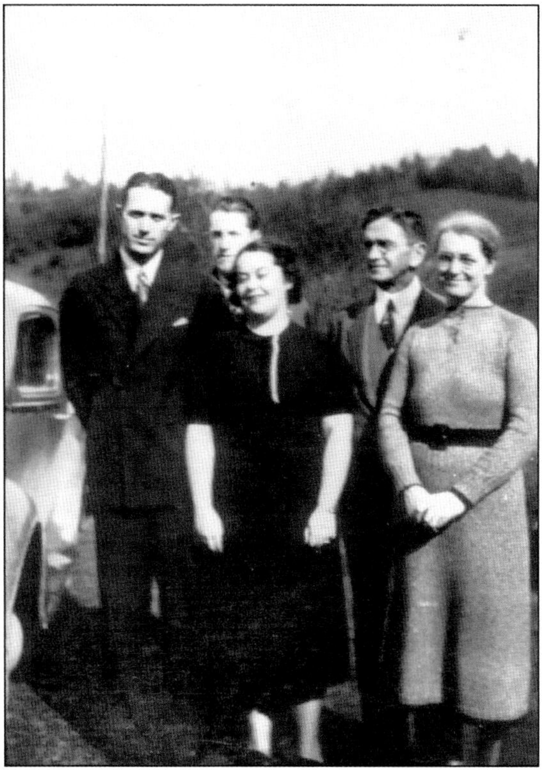

Pictured here, from left to right, are Raymond Coppenger, pastor of Butler Baptist Church; Robert DeVault, basketball coach and teacher of Watauga Academy; Helen K. Richie, first-grade teacher at Butler City School; Robert DeVault Sr., and Mrs. Robert (Bess) DeVault, principal of Butler City School. (Photograph courtesy of Butler Museum.)

Included in this photograph of the freshman class of 1936 of Watauga Academy are Jacqueline Farthing, Dora Grindstaff, ? Matherly, Kate Reece, Ester Arney, Velma Arney, Frances Courtner, Edwin DeVault, Coolidge Tester, Wayne Eggers, Sandy Greenwell, J. S. Stout, Estell "Pete" McQueen, Roy Ward, Ernest Snyder, C. L. Evans, Ruth Grindstaff, and Robert DeVault Jr. (Photograph courtesy of Mary Walker Ward.)

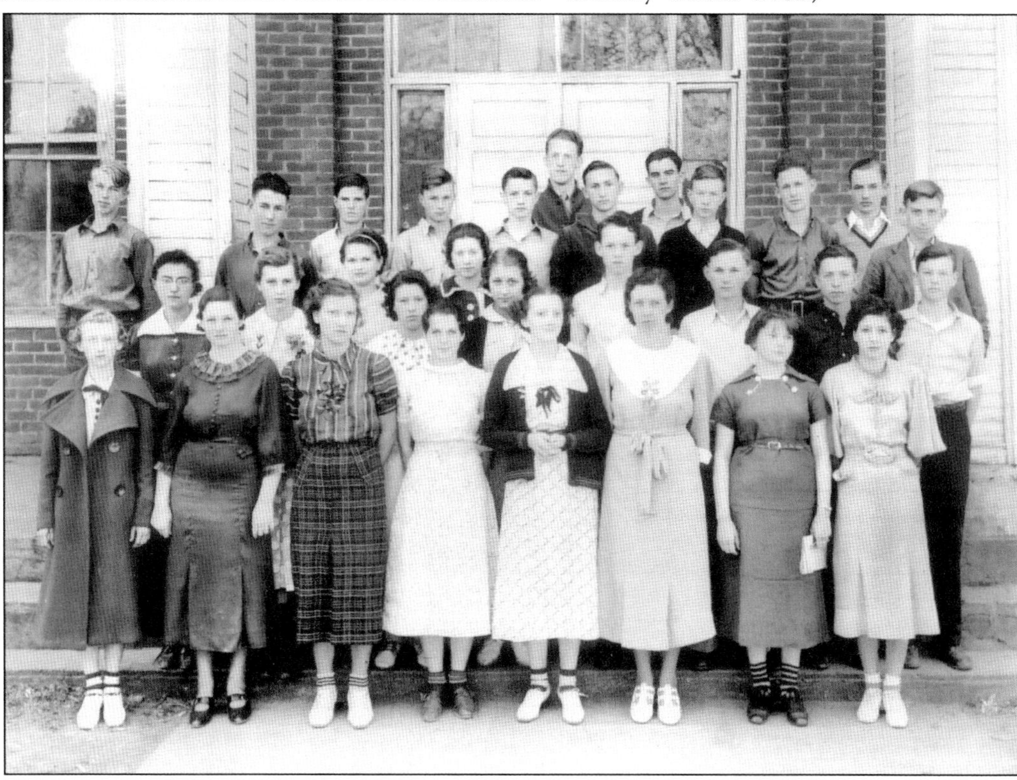

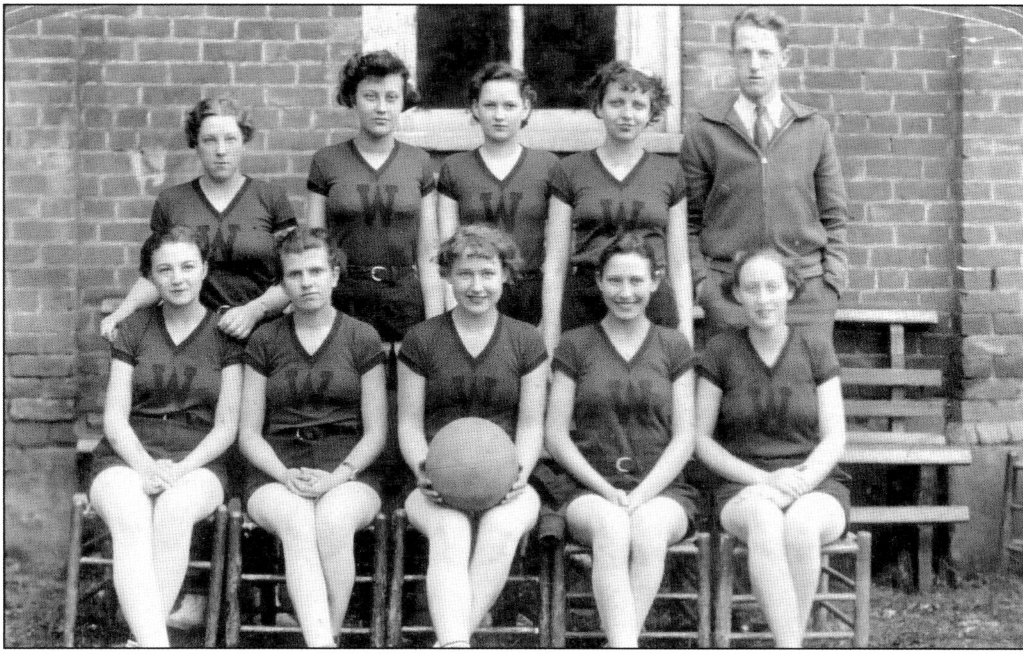

During the 1937–1938 school year, the Watauga Academy girls' basketball team included, from left to right, the following: (first row) Crystal McCloud, Buna McQueen, Leondas Farthing, Jane Proffitt, and Kate Smith; (second row) Marie VonCannon, Marie Laws, Lydia Cable, Margaret Hoghead, and Coach Robert DeVault Jr. (Photograph courtesy of Patsy Tadlock and Ermine Arwood.)

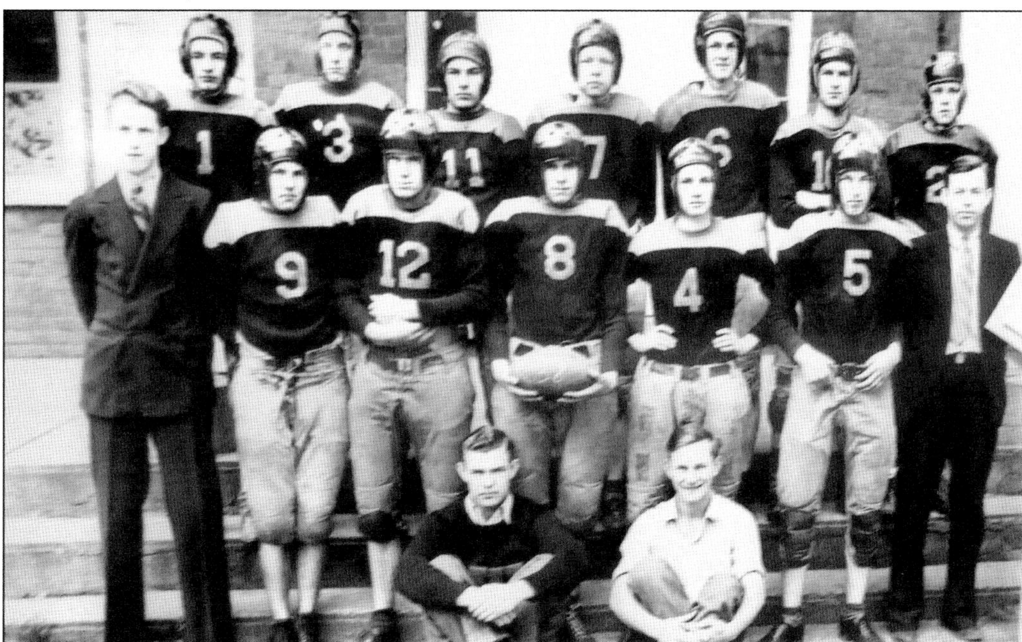

This 1938 photograph depicts Mr. Robert DeVault Jr. (far left) and Mr. Gordan Slemp (far right). They were coaches of the first football team in Watauga Academy's history. (Photograph courtesy of Butler Museum.)

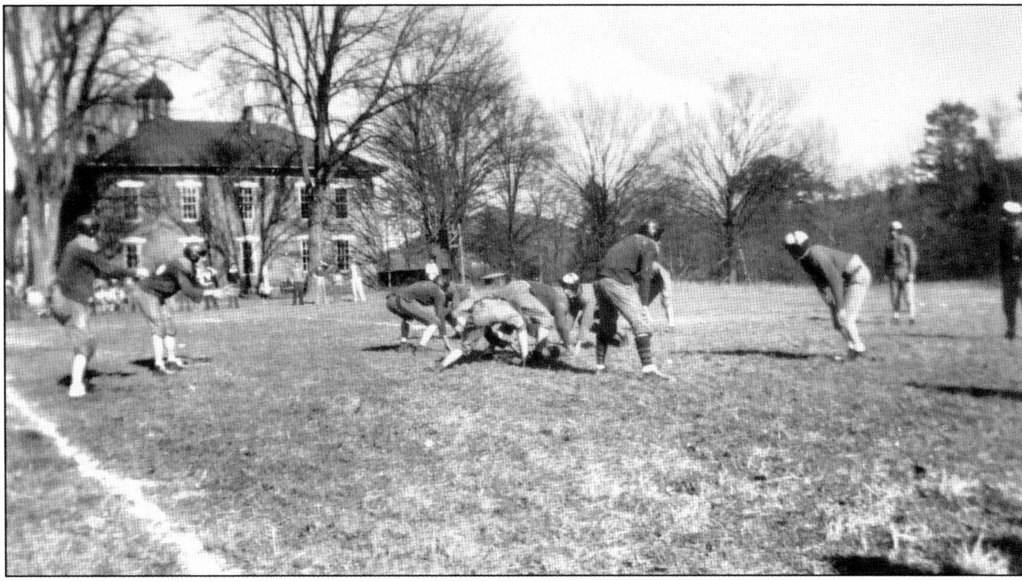

Students play football on the campus of the Watauga Academy in this image. (Photograph courtesy of Butler Museum.)

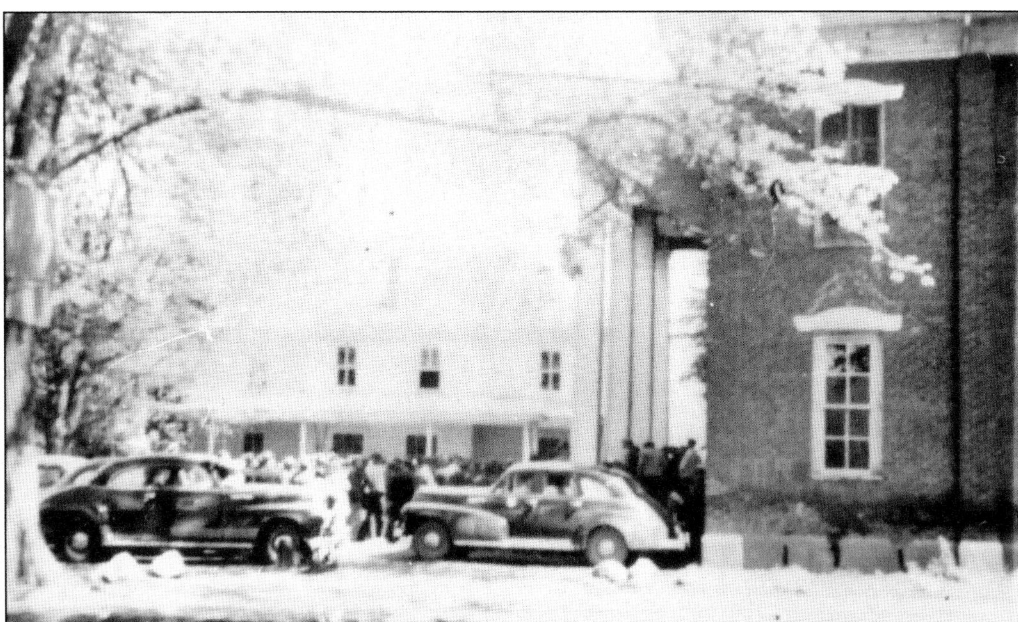

This photograph depicts some type of gathering at Watauga Academy during the late 1940s. Farthing Hall is in the background. Originally used as a girls' dormitory, Farthing Hall became the home economics building during the 1940s. (Photograph courtesy of Butler Museum.)

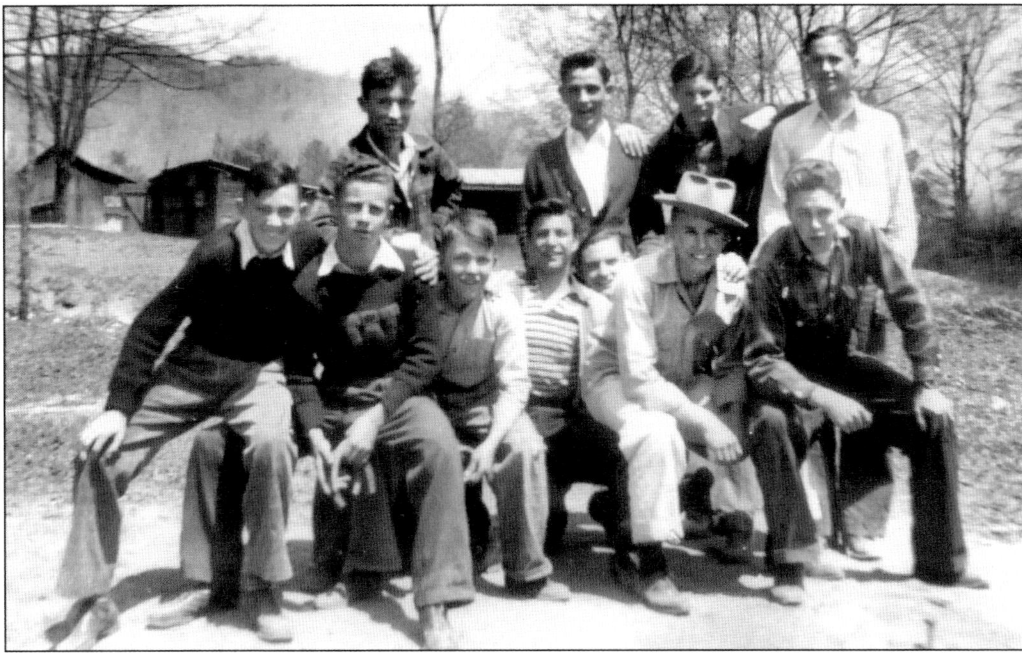

Included in this photograph, taken on the campus of Watauga Academy, are George Walker, Ray Trivette, R. D. Campbell, and Vaughn Dean. (Photograph courtesy of Butler Museum.)

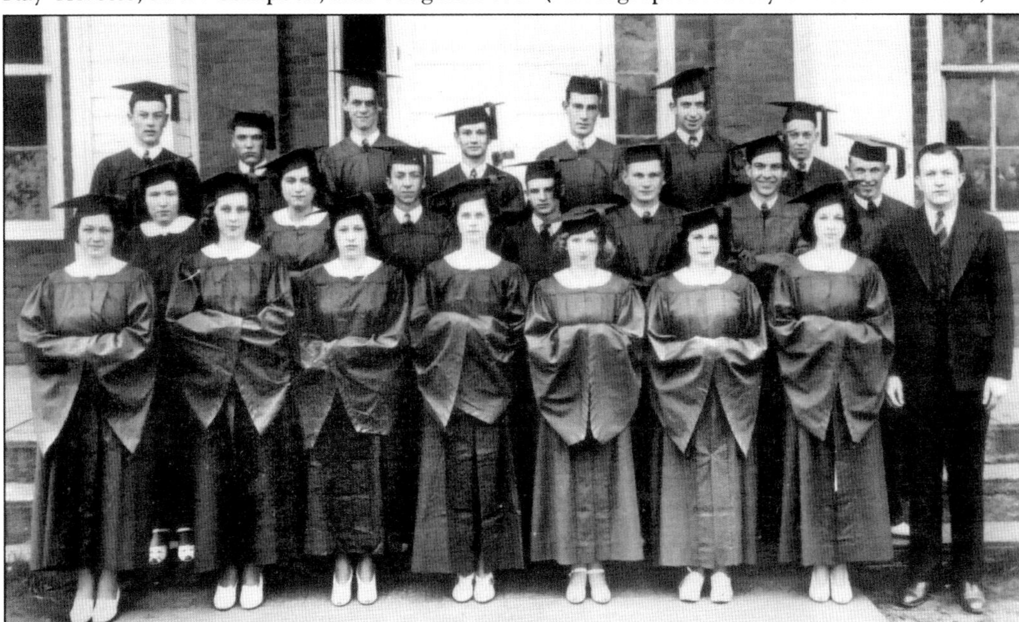

Graduates of the class of 1940 of Watauga Academy included the following: Warren Rainbolt, Velma Arney, Kate Reece, Frances Courtner, Edwin DeVault, Haynes Dugger, Wayne Eggers, C. L. Evans Jr., Jacqueline Farthing, Sandy Greenwell Jr., Dora Grindstaff, Ruth Harmon, Anna Holloway, Hazel McDonald, Earl Norris, Coolidge Tester, Elliot Thomas, Roy Ward, Grady Watson, Pauline Snyder, and J. S. Stout Jr. The principal at far right in the photograph is Paul Cates. (Photograph courtesy of Mary Walker Ward.)

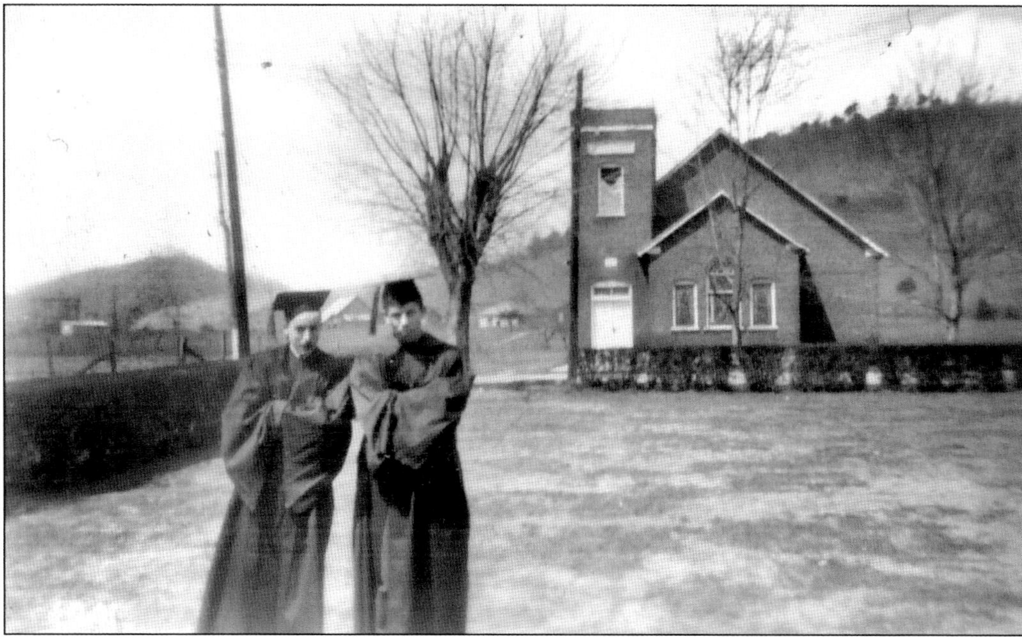

This photograph of J. L. Evans (left) and Cleve Clawson was taken in 1941 upon their eighth-grade graduation from Butler City Schools. Both young men lost their lives during World War II. (Photograph courtesy of Butler Museum.)

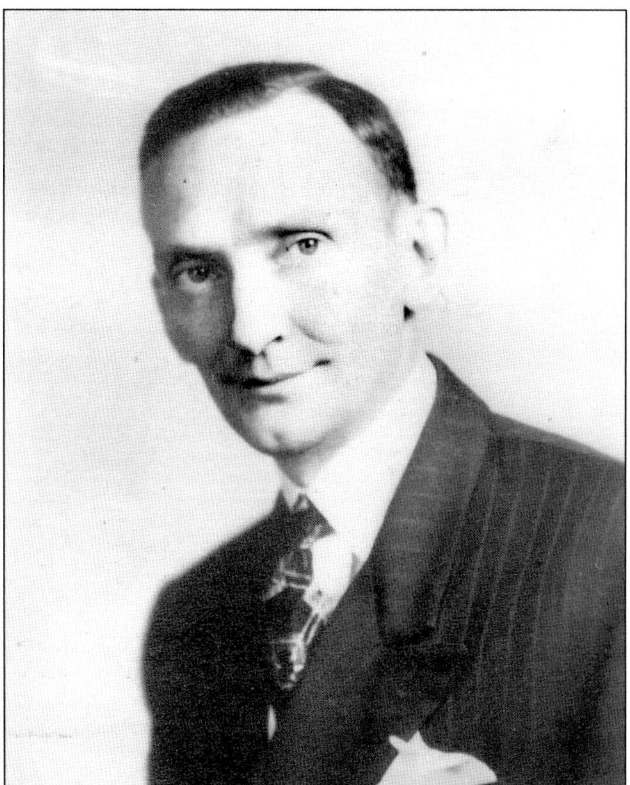

George H. Jenkins served as principal of Watauga Academy from 1943 to 1945. (Photograph courtesy of Butler Museum.)

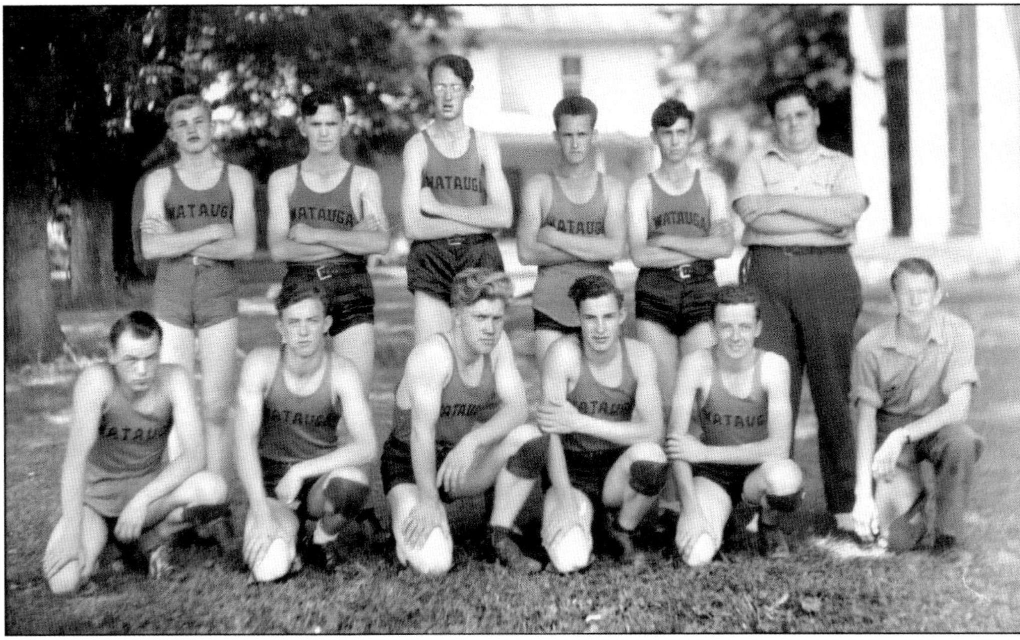

The basketball team of Watauga Academy during the 1943–1944 school year included, from left to right, the following: (first row) Mack Wolf Jr., George McCloud, Sherman Walsh, James Jenkins, Bill Trivette, and Charles Jenkins (manager); (second row) Charles "Bill" Jenkins, Johnny Proffitt, Ralph Hyder, Clyde Watts, Paul Courtner, and Coach Haven Lowe. (Photograph courtesy of Mary Walker Ward.)

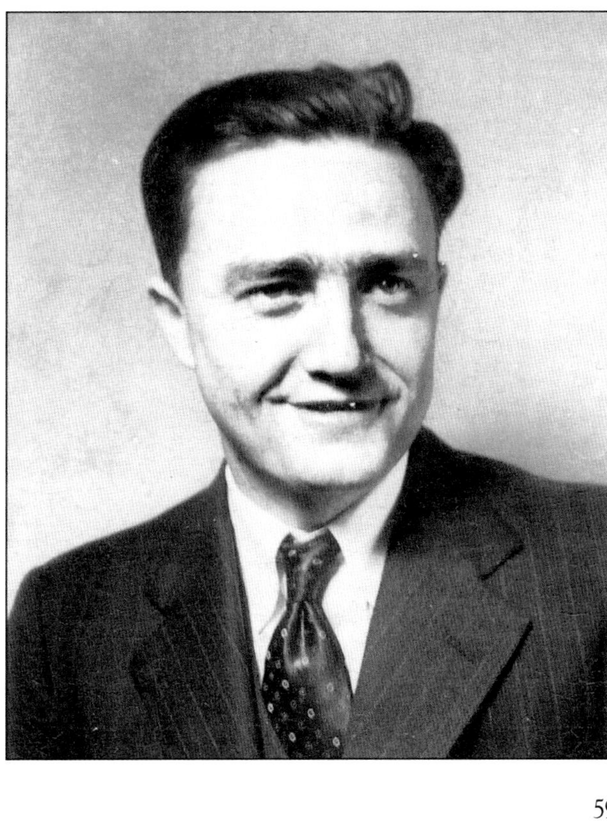

Rev. Coy Riddle served as principal of Watauga Academy and pastor of Butler Baptist Church during the mid-1940s. (Photograph courtesy of Butler Museum.)

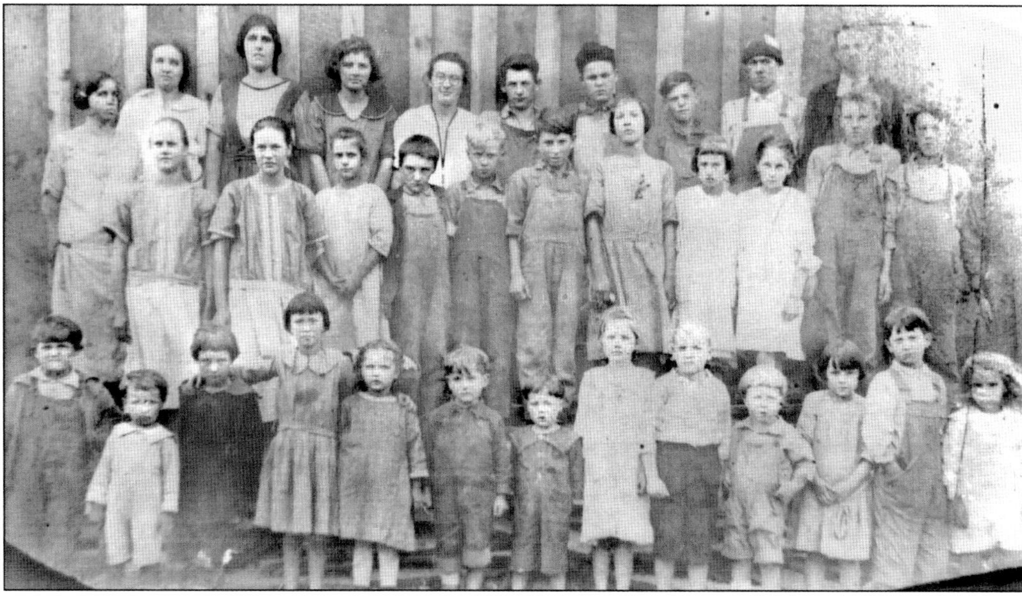

This 1925 photograph of students at Frog Pond School includes, from left to right, the following: (first row) Fred Dugger, Hillard Triplett, Faye Triplett Isaacs, Rosa Bunton Baker, Lois Cable, Clint Lunceford, Clay Cable, Crete Cable Payne, David Cable, Floyd Gilbert, Fannie Bunton Baker, Howard Pearson, Milda Isaacs Stout, and Nannie Dugger; (second row) Eliza Cable Simerly, Judy Cable Hobson, Rosa Cable Campbell, Turner Triplett, Haskel Isaacs, Elmer Smith, Bonnie Bowman, Ersel Bunton Wolfe Tester, Lucy Cable Smith, Tom Bunton, Clyde Holden, and Don Bowman; (third row) Mae Reece Lewis, Martha Lunceford Bunton, Sally Cable Bunton, Jane Lunceford Holden, Flora Lunceford Culberson Myers, Tine Slemp, Stacy Lunceford, Tom Holden, Tal Bowman, and Tip Bunton. Frog Pond School washed away when the Watauga River flooded in 1940. (Photograph courtesy of Juanita Tester Wilson.)

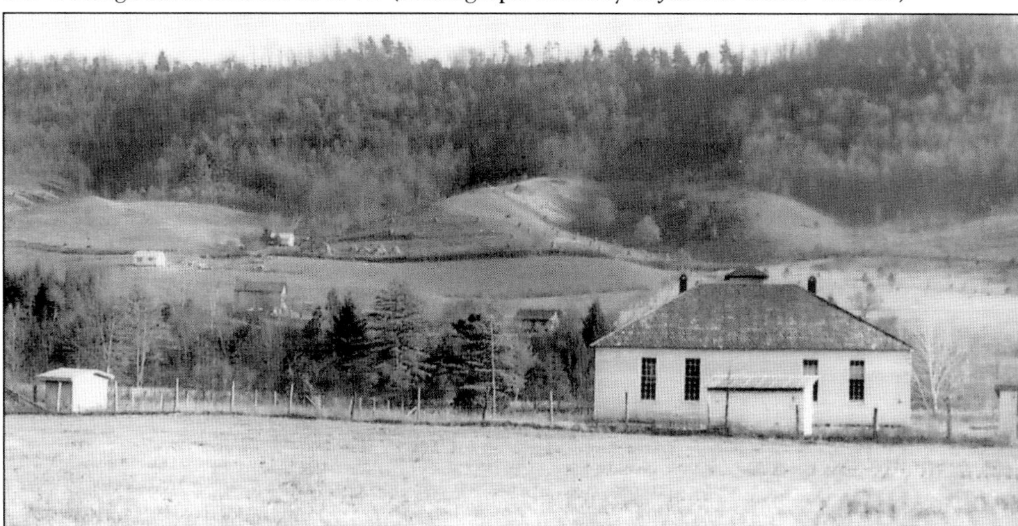

The student population at Rock Springs School, located in Doeville, declined long before the Watauga Reservoir was flooded. The lumber business, a primary source of income for many families in the area, practically came to an end following the flood of 1940, which washed out much of the Southern Railroad. (TVA photograph courtesy of Butler Museum.)

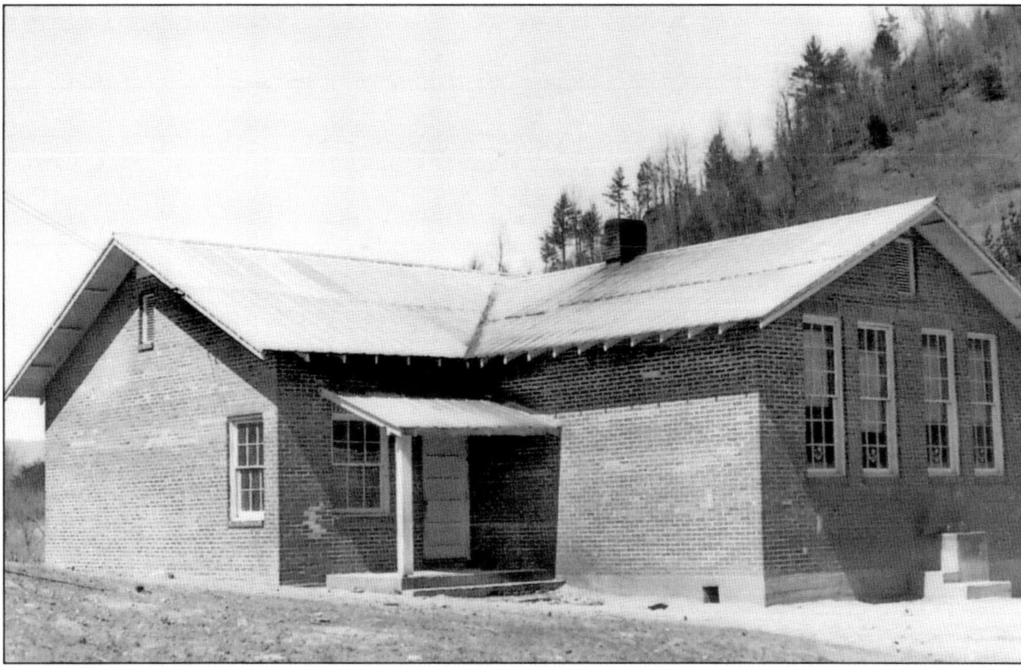

Located in the Elk River community, the Lower Elk Elementary School building was built in 1937–1938 by Jobe Eggars and his sons. Rather than relocating out of the Watauga Reservoir, the building was torn down in the 1940s. Students were transferred to nearby schools at Elk Mills and Little Milligan. (TVA photograph courtesy of Butler Museum.)

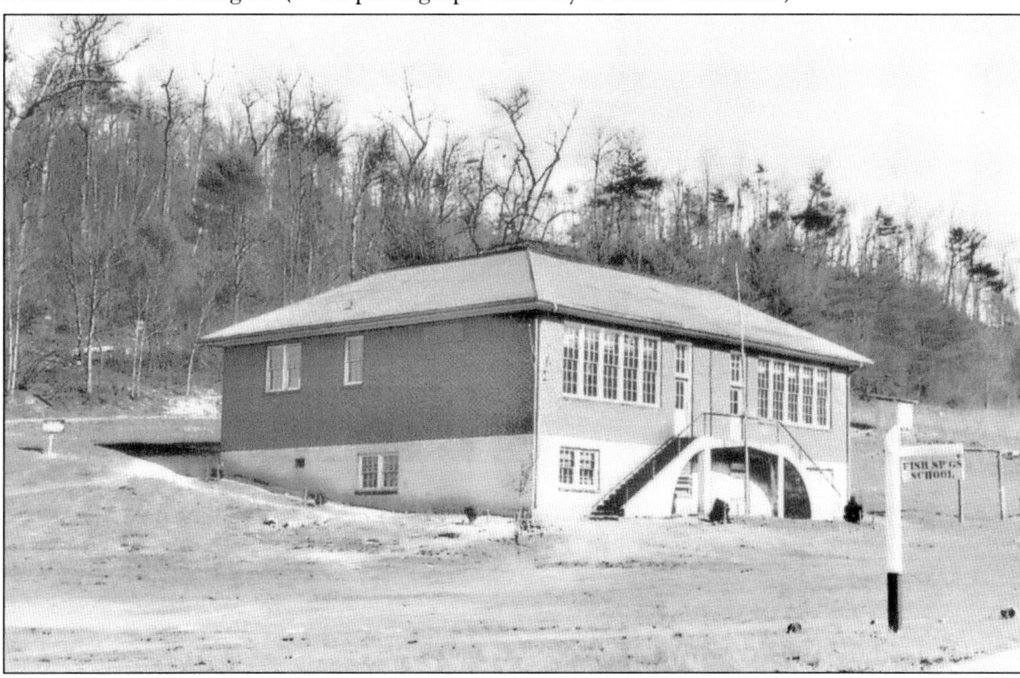

Erected in 1917, Fish Springs School served the families that lived between Hampton and Butler for many years. Upon its closing, the students were transferred to nearby schools at Hampton and Little Milligan. (TVA photograph courtesy of Butler Museum.)

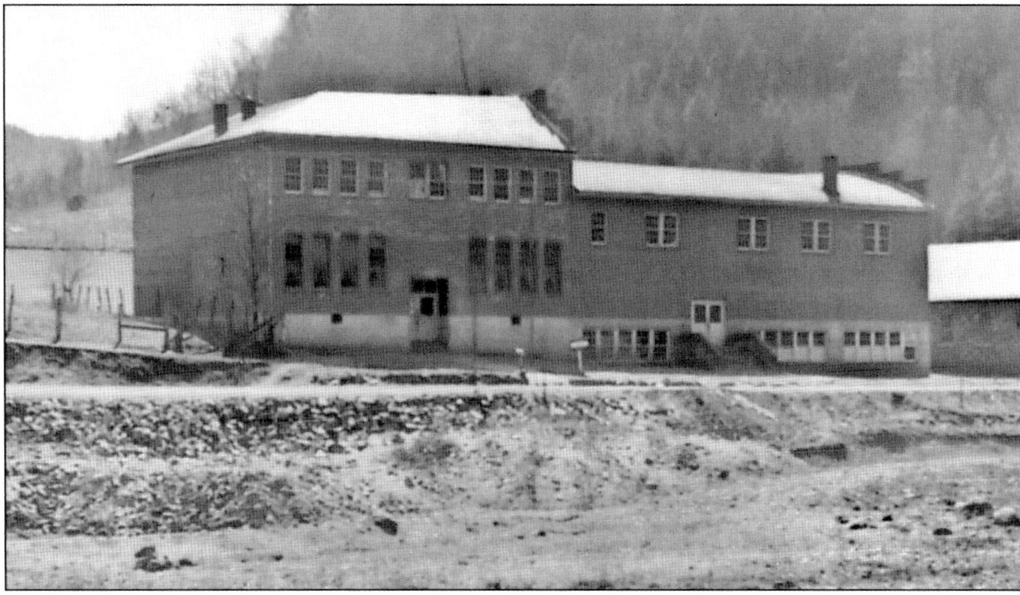

Located in the Elk Mills community, the Elk Mills School served the area for several years before closing in early 1960. Students were transferred to nearby schools at Little Milligan and Hampton. (TVA photograph courtesy of Butler Museum.)

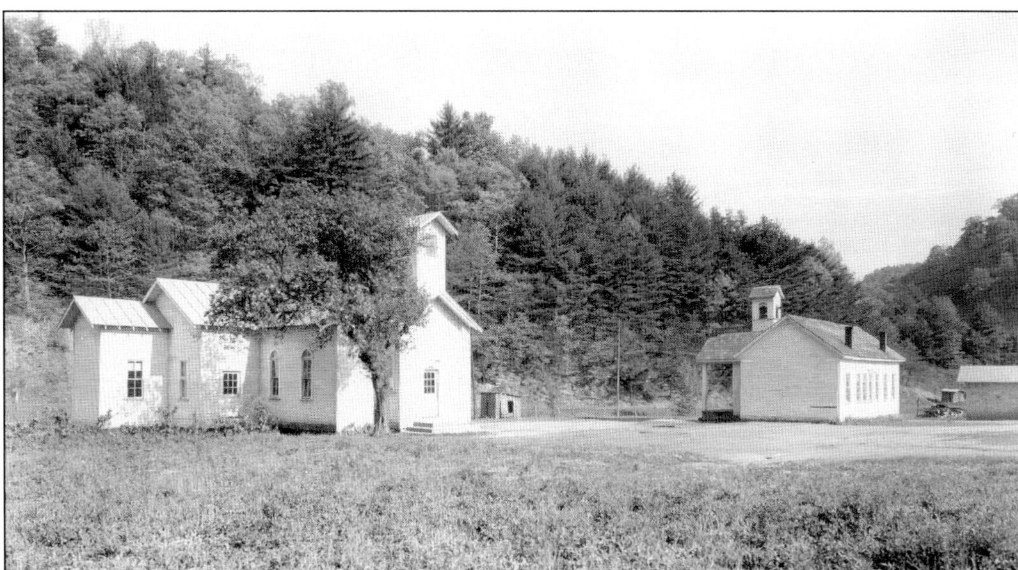

Cobbs Creek School was located in the Cobbs Creek community. In 1948, Cobbs Creek School and Butler City School combined to form the Watauga Elementary School in Carderview (later known as Butler). (Photograph courtesy of Butler Museum.)

Four

CHURCHES

As the businesses and schools developed in the Watauga Valley, churches began to dot the countryside. These churches aided the spiritual development of the communities as they grew and prospered materially. A number of different denominations were represented in the Watauga Valley, reflecting the diverse beliefs in the communities. Yet in spite of these theological differences, these congregations focused on the things they held in common. As with the schools, the fate of these churches varied when the Watauga Dam was built. Some relocated along with their members, some were untouched by the rising waters but struggled when their members were forced to relocate, while others were "disbanded and dismantled" and their members scattered to other congregations in the region.

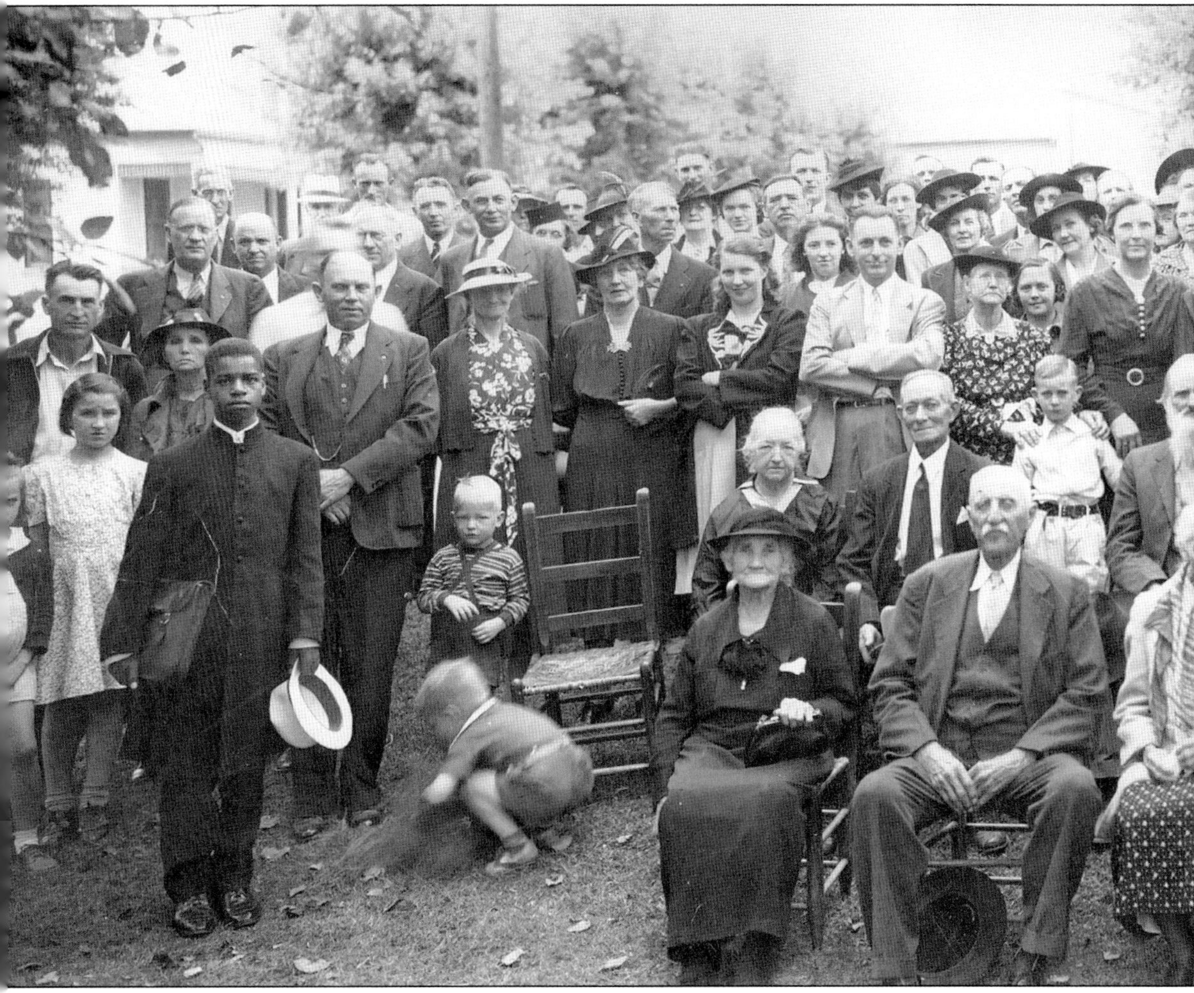

In 1939, members of many congregations from the different denominations in the area gathered together to attend a revival at the Butler Methodist Church. Reverend Breedlove was the

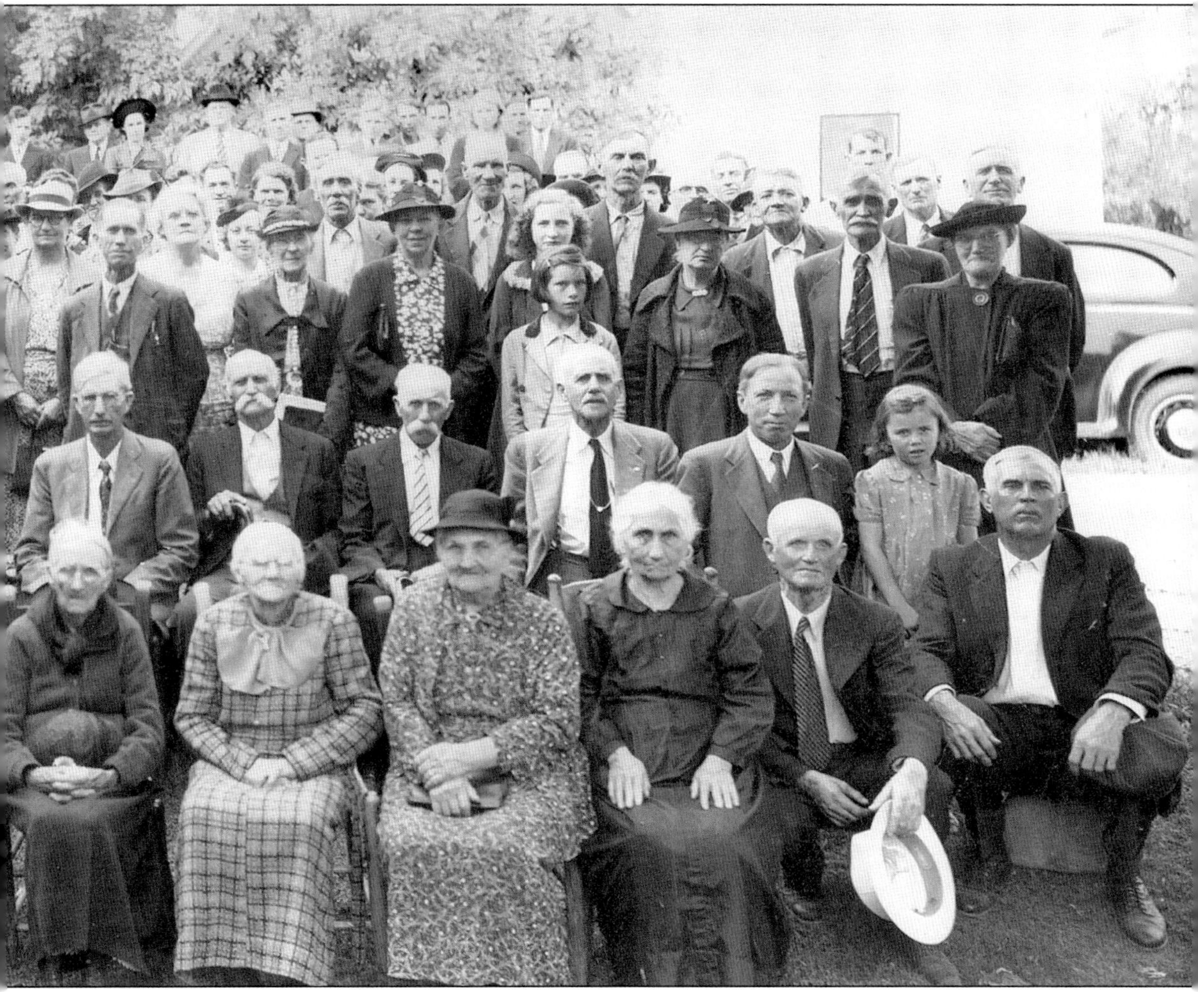

visiting minister. (Photograph courtesy of Butler Museum.)

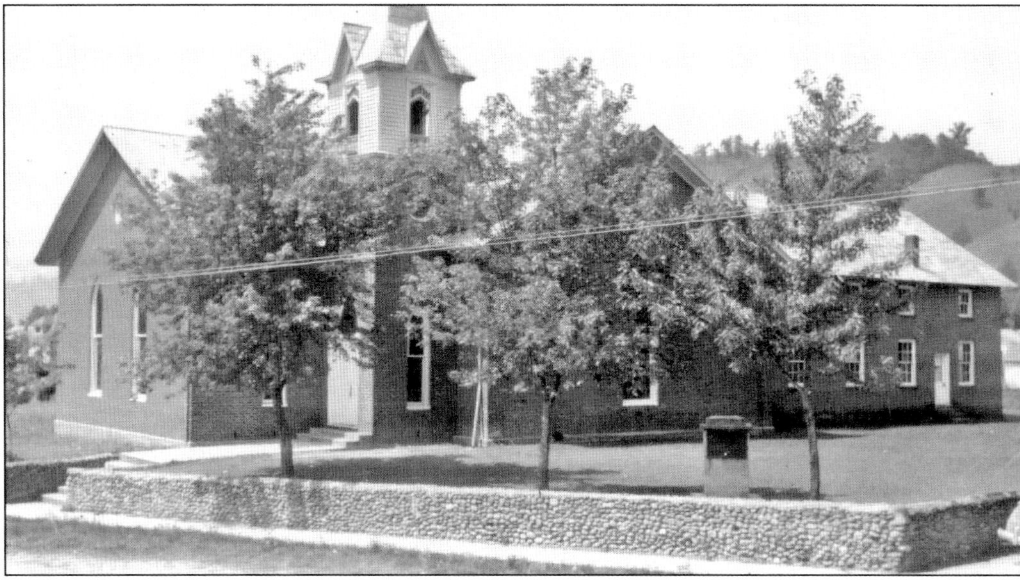

Once known as Holly Spring Baptist Church, the Butler Baptist Church has served the Butler and Cobbs Creek communities since the late 1800s. When the Watauga Dam was built, the church's property located on Main Street was sold and the educational wing of the building was relocated to the church's present site in New Butler (then known as Carderview). This photograph of Butler Baptist Church was taken by pastor Raymond A. Coppenger in 1939. Notice the Daniel Boone marker erected in 1914. (Photograph courtesy of Butler Museum.)

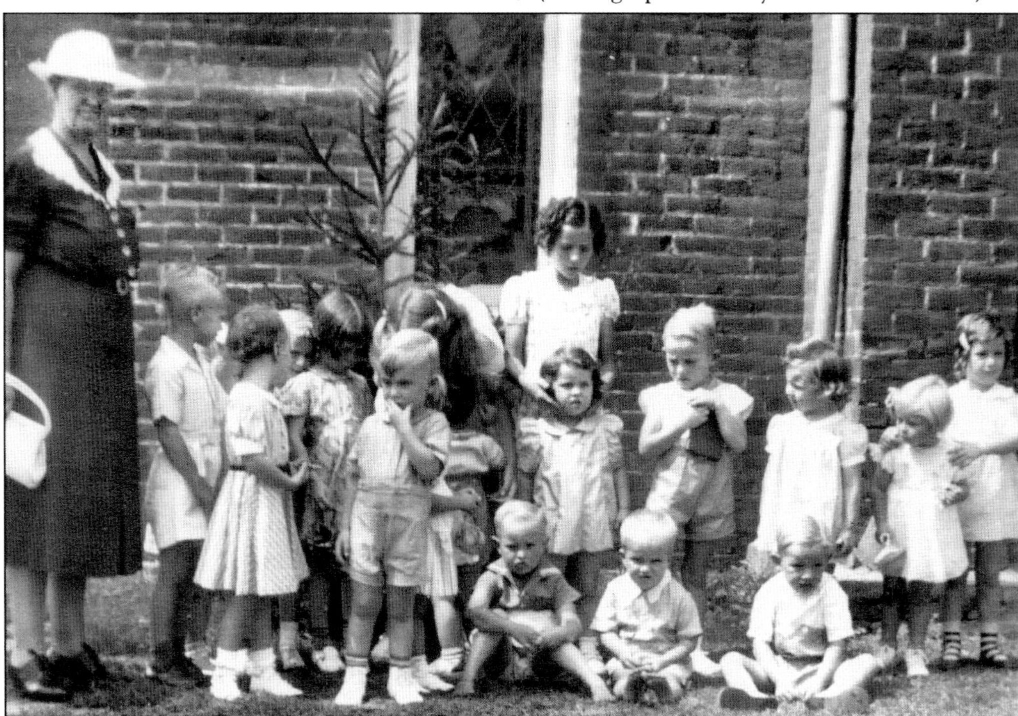

This photograph, taken in 1940, is the cradle-roll class of Butler Baptist Church, taught by Mrs. Lona Smith. (Photograph courtesy of Butler Museum.)

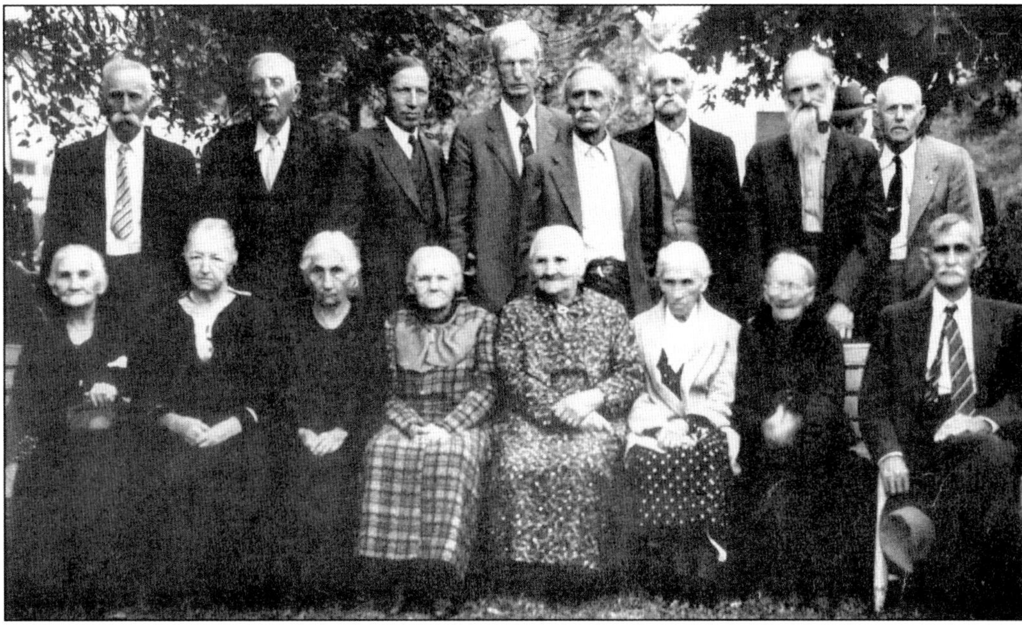

The congregation of the Butler Baptist Church, from left to right, includes the following: (first row) Mrs. David Whitehead, Mrs. Grant Ellis, Mrs. Lee Bailey, Mrs. Nancy Hazelwood, Mrs. Wilburn Reece, Mrs. Fronia Rainbolt, Mrs. Nancy Vines, and Grant Ellis; (second row) John Smith, W. O. Phillips, B. Carroll Reece, Joe H. Gregg, Ensor McNeal, David Whitehead, John Cable, and W. Y. Simerly. (Photograph courtesy of Butler Museum.)

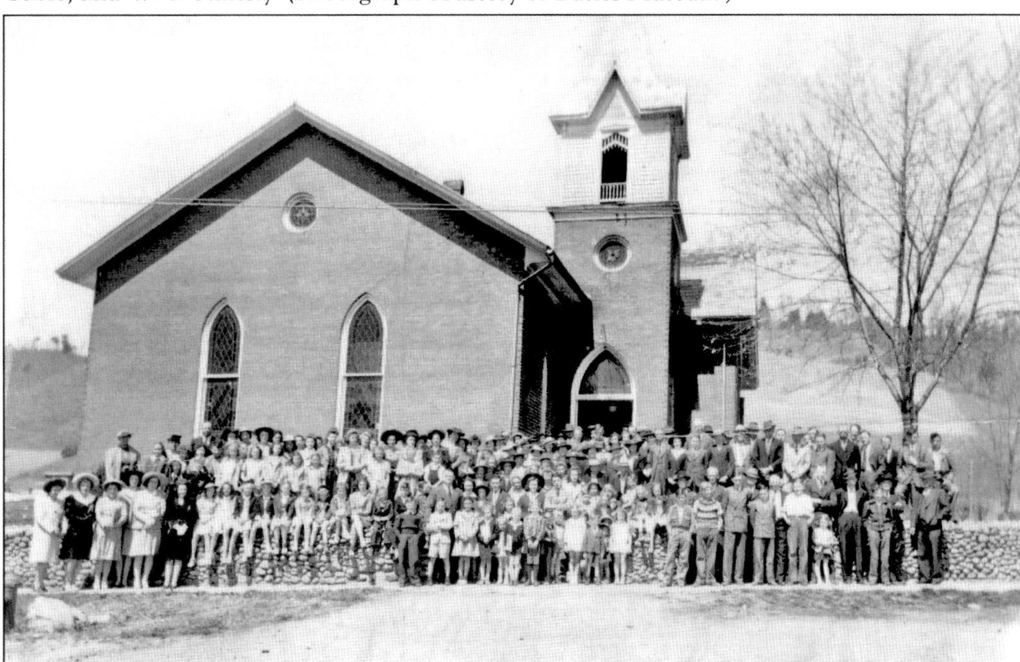

Pastor James Gregg (second row, center) stands among the congregation of the Butler Baptist Church. This photograph was taken during the early 1940s, just before the young men of the congregation left for military service. (Photograph courtesy of Butler Museum.)

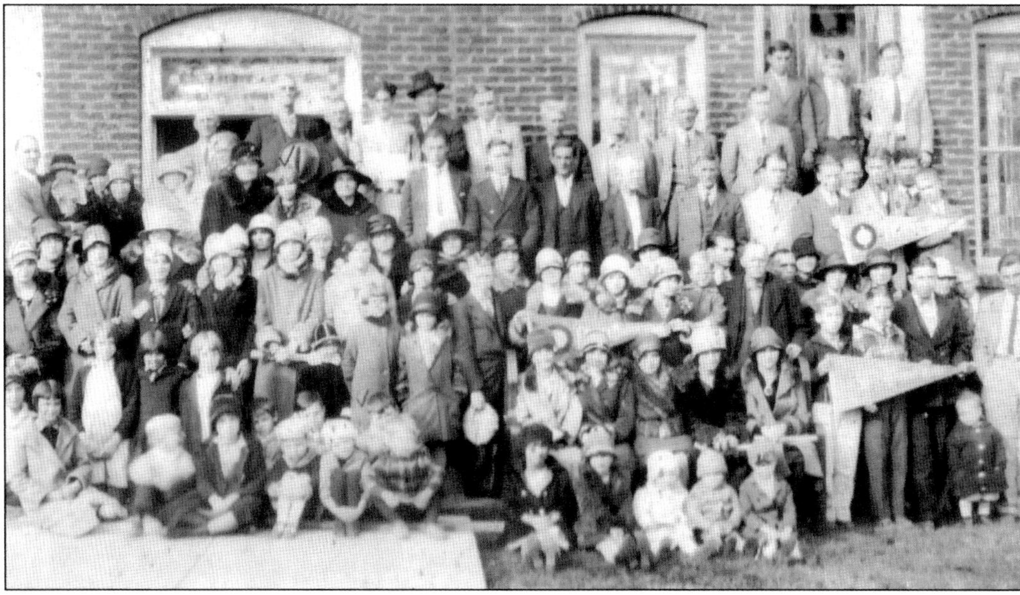

Pictured among the congregation of Butler Christian Church are B. A. Lipford Jr., Pauline Lipford, and Kay Ritchie. The photograph was taken in 1923. (Photograph courtesy of Mary Lou Gross.)

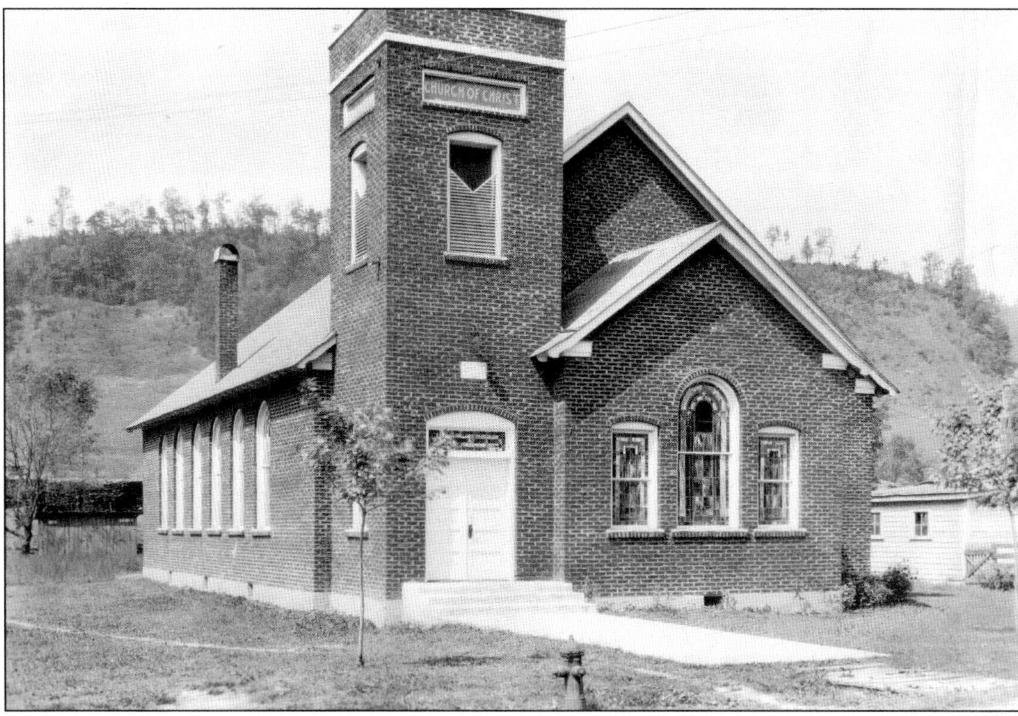

Butler Christian Church served the community until its members were forced to relocate out of the Watauga Reservoir in the late 1940s. Located on College Street, the church building was dismantled instead of relocated. Its bricks were used to brick two houses in Hampton and the furniture was distributed to several other churches. Notice the fire hydrant in the photograph, demonstrating the advanced development of the town despite its isolated location. (Photograph courtesy of Butler Museum.)

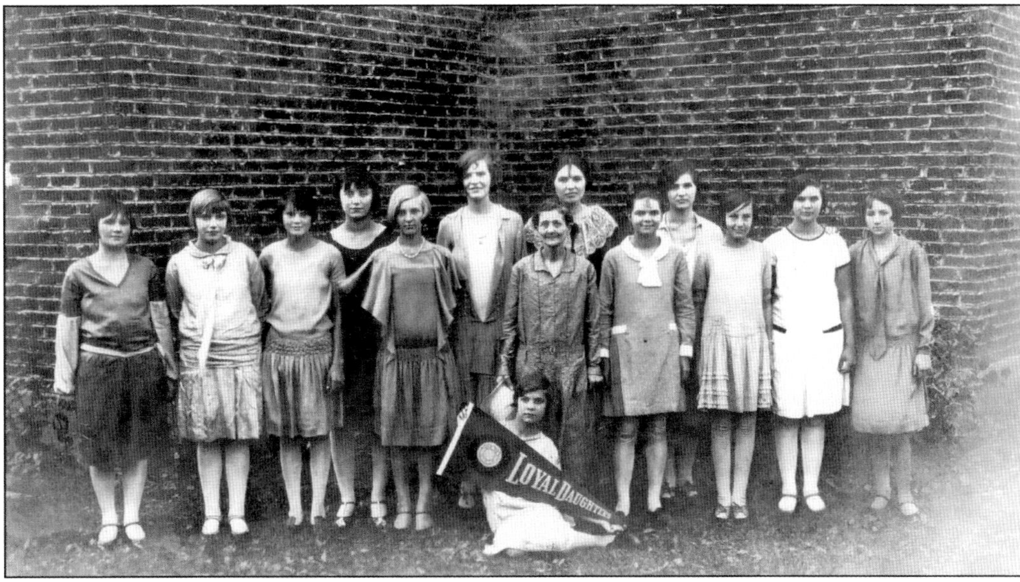

Included in this image of the Loyal Daughters, a Christian organization of girls, are, from left to right, Mae Whitehead, Selma Stout Wilson, Mrs. Flora Greene, Vivian Reese, Gladys Wilson, Nelle Donelly, Irene Shoun, Crete Netherly Wilson, Nelle Leonard, and Inez Leonard. The photograph was taken near Butler City School. (Photograph courtesy of Butler Museum.)

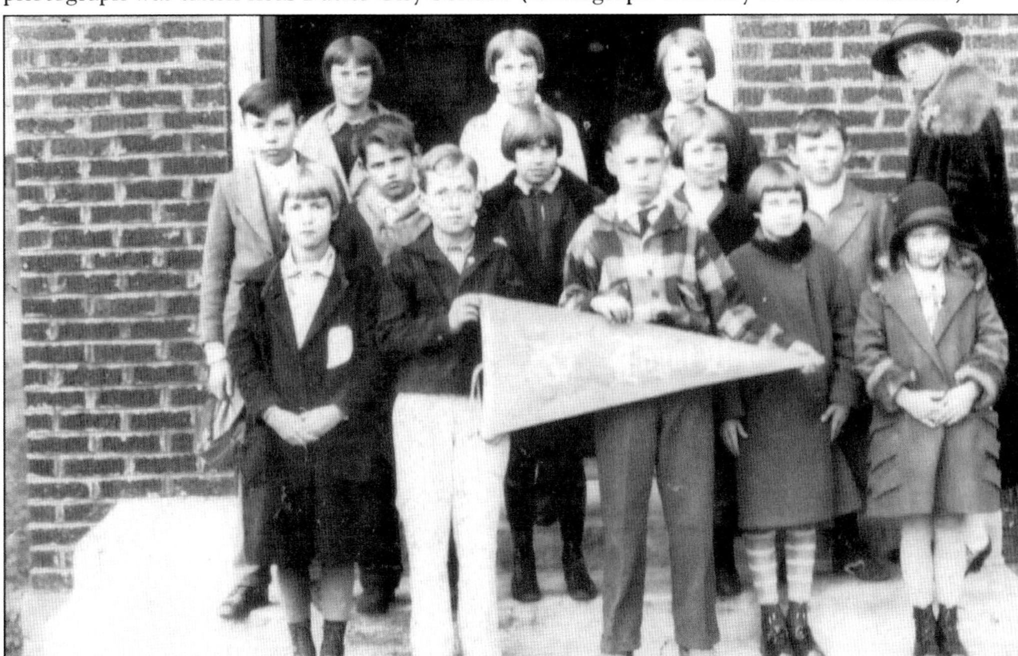

Butler Christian Church was organized in 1912 with Mr. Clarence Poage serving as the first minister. Shown in this photograph are, from left to right, the following: (first row) Ruby Griffey, Frank Whitehead, Vernon Wilson, unidentified, and Ruth Lewis; (second row) French Neatherly, Ray Wilson, unidentified, Chastine Griffey, and B. A. Lipford Jr.; (third row) Hazel Donnelly, Dorothy Whitehead, and Ruth Griffey. Mrs. Wheeler Wilson served as a teacher. (Photograph courtesy of Butler Museum.)

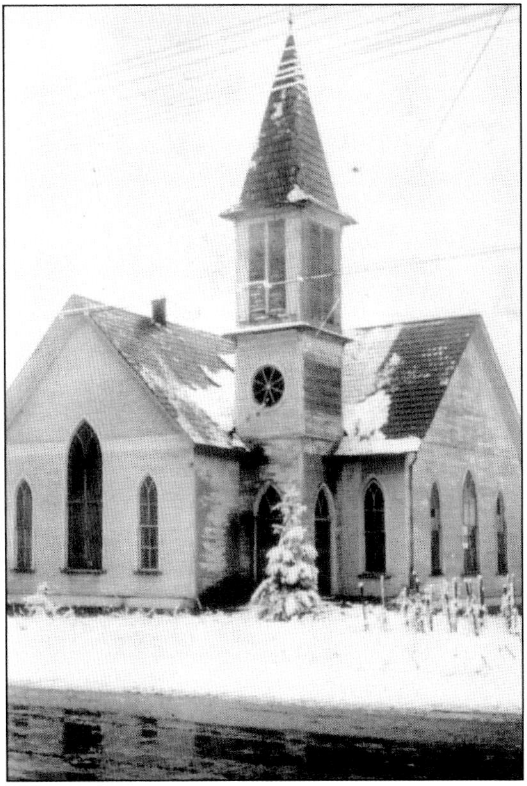

Located on Spring Street (also called Depot Street), Butler Methodist Church was active from 1911 until 1948. When the Watauga Dam was built, the property was sold and church membership was transferred to other churches as families relocated out of the Watauga Reservoir. J. T. Ware served as the church's first minister. (Photograph courtesy of Butler Museum.)

This photograph of the Cobbs Creek Baptist Church congregation was taken *c.* 1938. During the late 1940s, the church building was torn down and a new building was built in Carderview (later known as Butler). (Photograph courtesy of Butler Museum.)

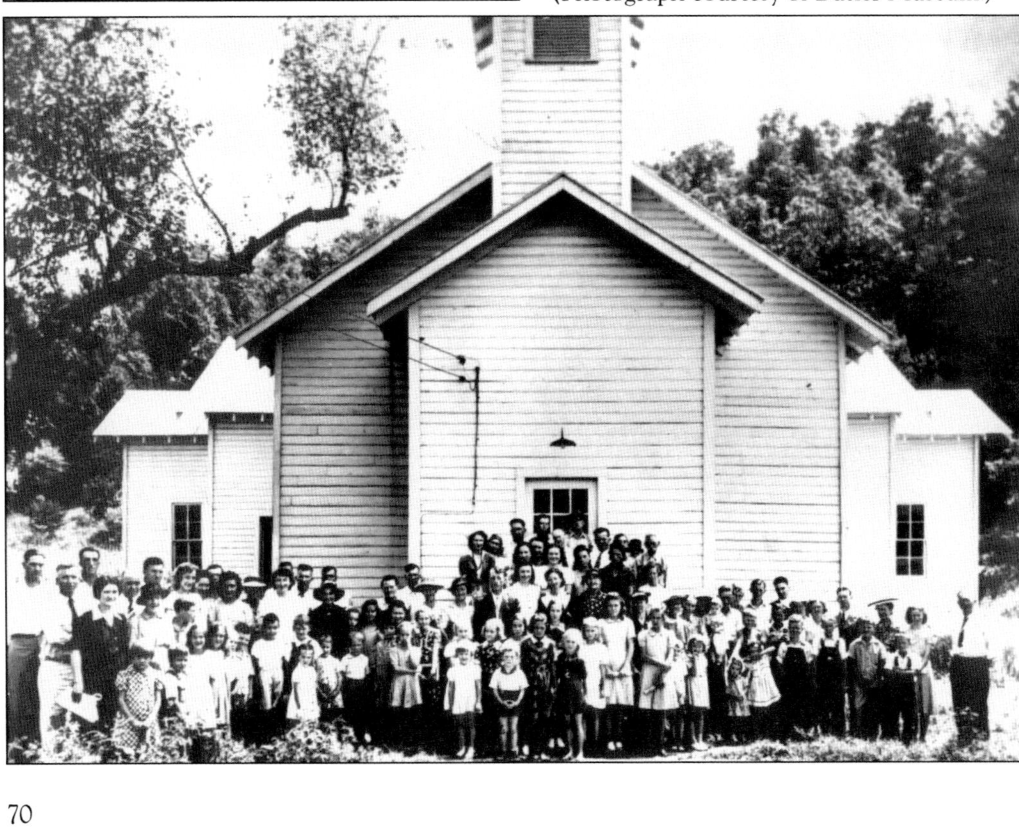

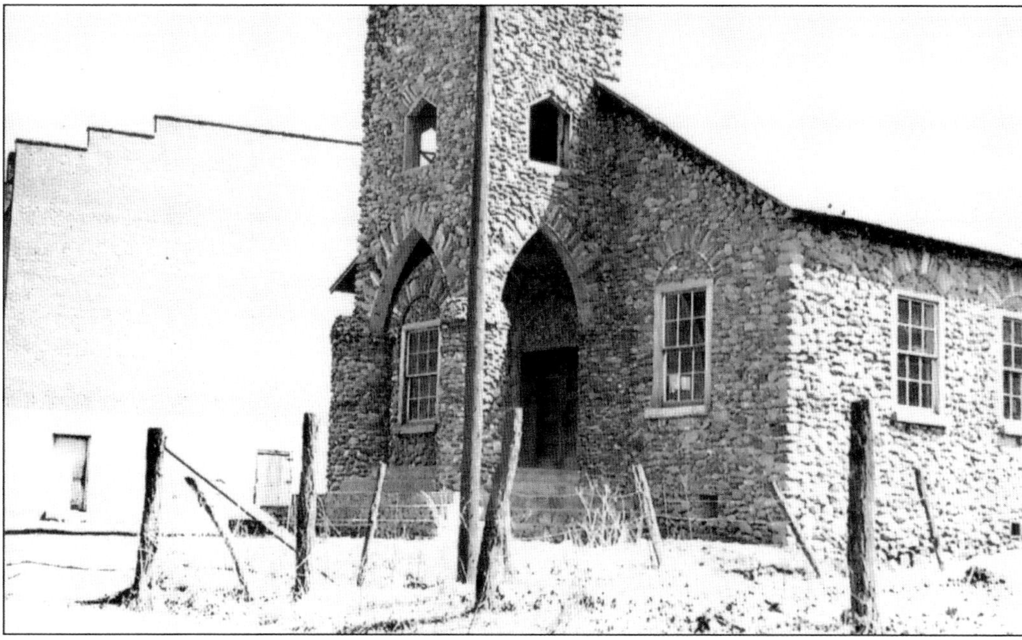

After sharing a building with the Elk Mills School and community for more than 30 years, the Elk Mills Christian Church constructed a new edifice during the 1930s. The building, adorned with stones harvested from Elk River, stood above the water line of the Watauga Reservoir and acquired the communion table and other furniture from the Butler Christian Church. (TVA photograph courtesy of Butler Museum.)

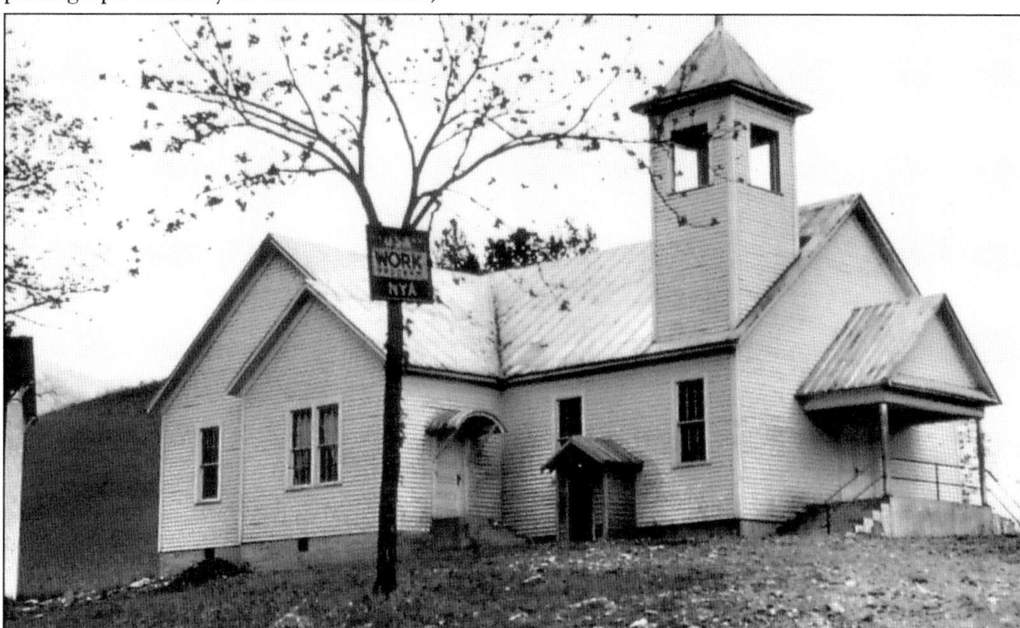

Union Baptist Church, a member of the Watauga Baptist Association since 1871, was located near the Watauga Dam site in Carden's Bluff. In 1943, the property was sold to the Tennessee Valley Authority and the building was moved out of the Watauga Reservoir. (TVA photograph courtesy of Butler Museum.)

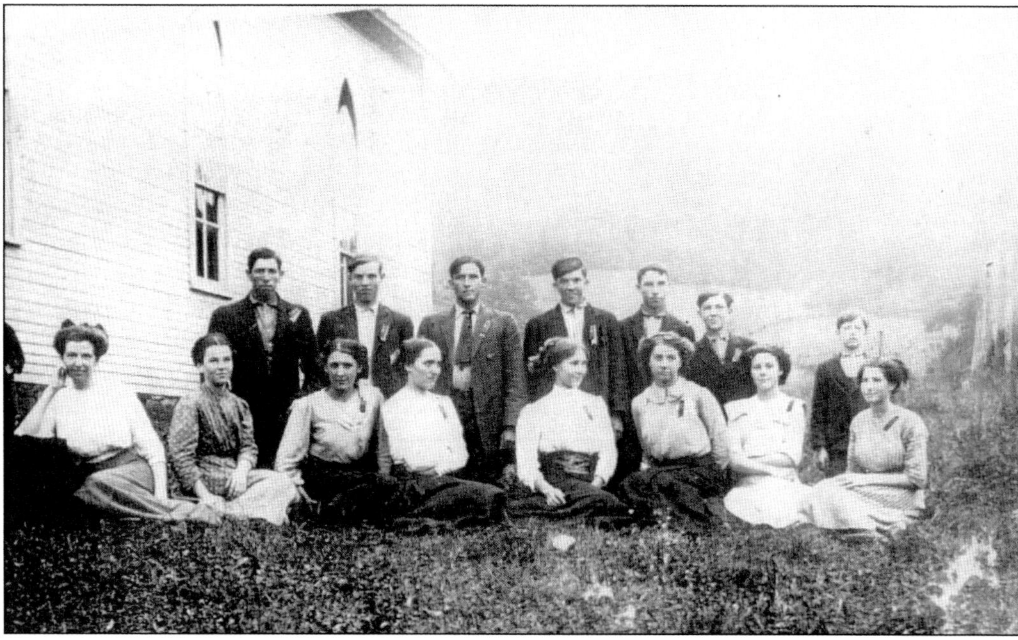

Shown in this photograph, taken at Sugar Grove Baptist Church, are the following from left to right: (first row) Elsie Crosswhite, Nettie Cable, China Phillips, Lula Forrester, Pollyanna Forrester, Lockine Dugger, Leckie Gregg, and Ethel Isaacs; (second row) Frank Isaacs, Roy Isaacs, Spurgeon Tester, Ira Dugger, Roy Dugger, Hunter Dugger, and Frank Dugger. (Photograph courtesy of Juanita Tester Wilson.)

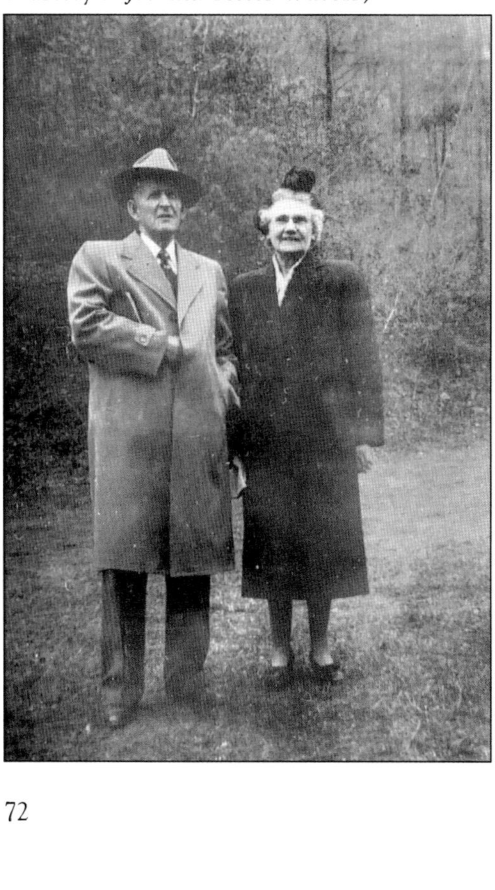

Rev. and Mrs. J. J. Richardson served Sugar Grove Baptist Church from 1934 to 1950. (Photograph courtesy of Juanita Tester Wilson.)

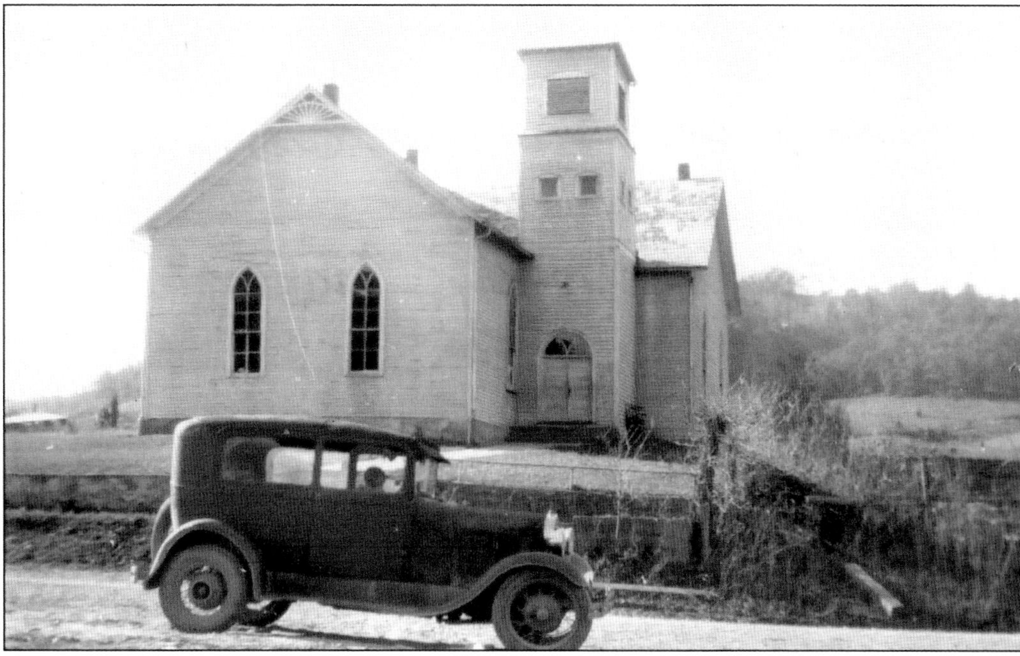

Originally named "The Baptist Church of Christ at Sugar Grove," the Sugar Grove Baptist Church was organized on July 12, 1850. In 1949, the church building and cemetery were moved out of the Watauga Reservoir onto land acquired from A. B. Wagner and J. H. Gregg. (Photograph courtesy of Juanita Tester Wilson.)

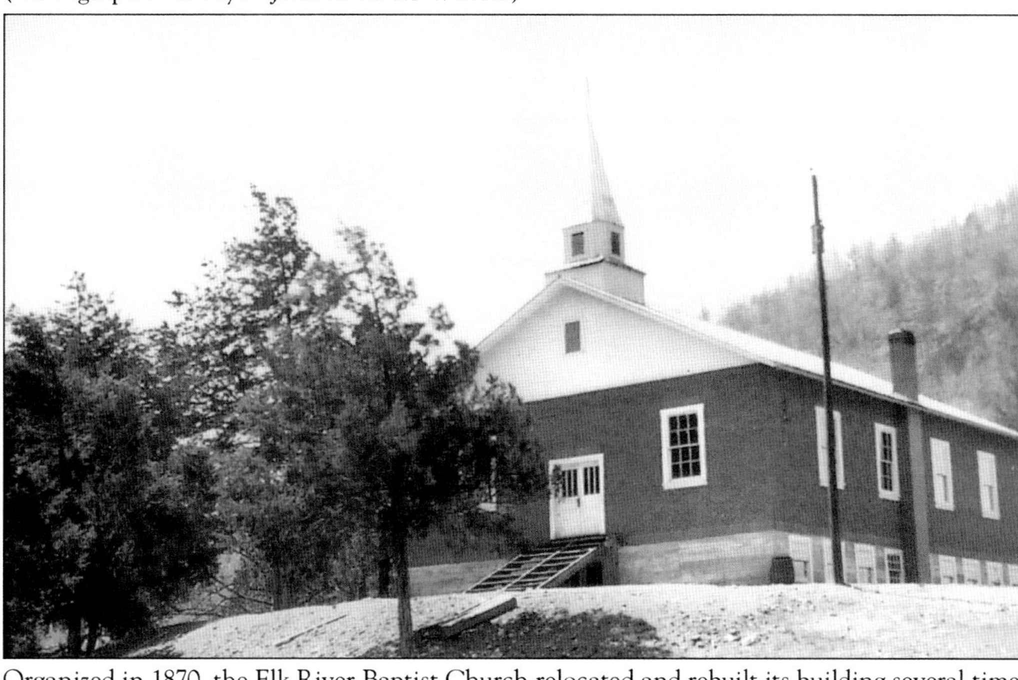

Organized in 1870, the Elk River Baptist Church relocated and rebuilt its building several times to accommodate the growth and needs of its congregation. However, in 1947, the church was forced to relocate once again—this time outside of the Watauga Reservoir. (TVA photograph courtesy of Butler Museum.)

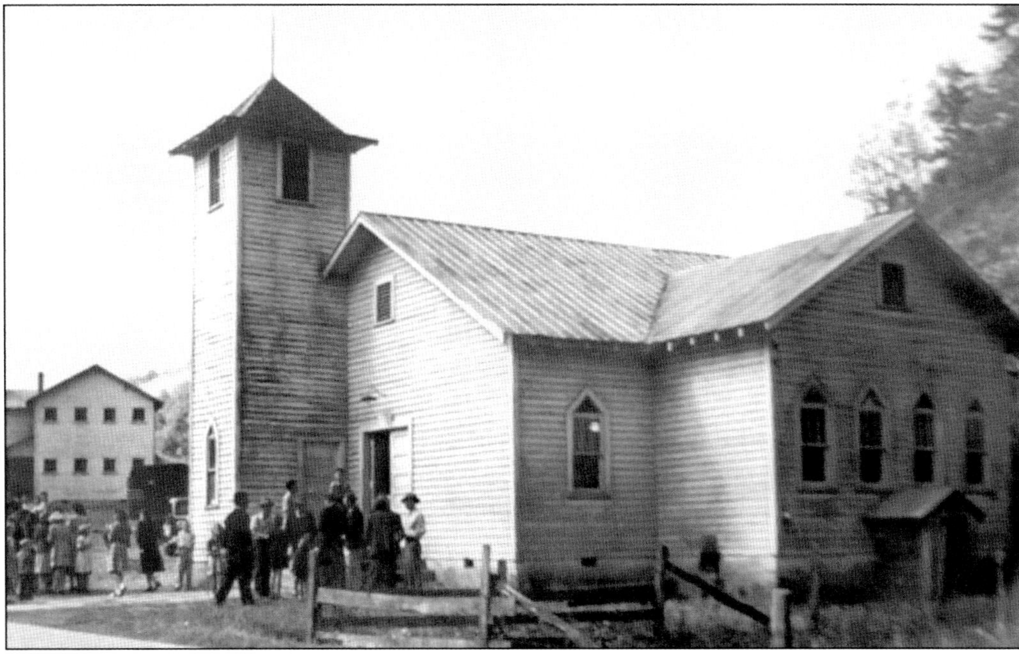

Established in 1903, the Little Milligan Baptist Church has served the Little Milligan community for more than 100 years. The large wooden building, which also served as a school until 1923, was moved out of the Watauga Reservoir in the late 1940s. W. Joe Potter, the first pastor of Little Milligan Baptist Church, served the congregation for more than 15 years. (Photograph courtesy of Butler Museum.)

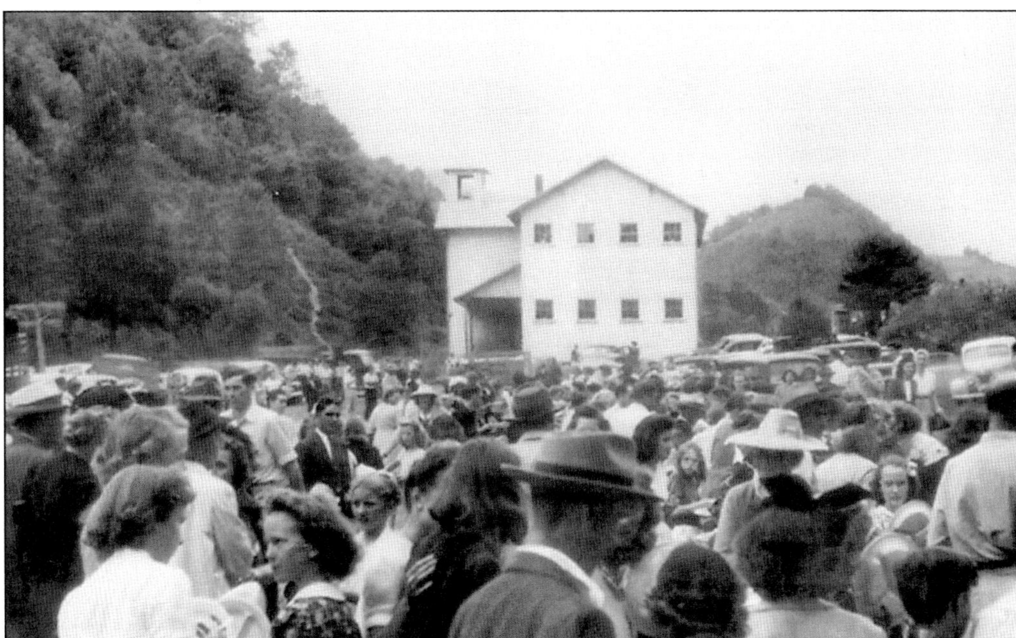

Each year, the Watauga Baptist Association would celebrate its homecoming at a church in its membership. This picture depicts a gathering held at Little Milligan Baptist Church, which may have been one of those homecomings. (Photograph courtesy of Butler Museum.)

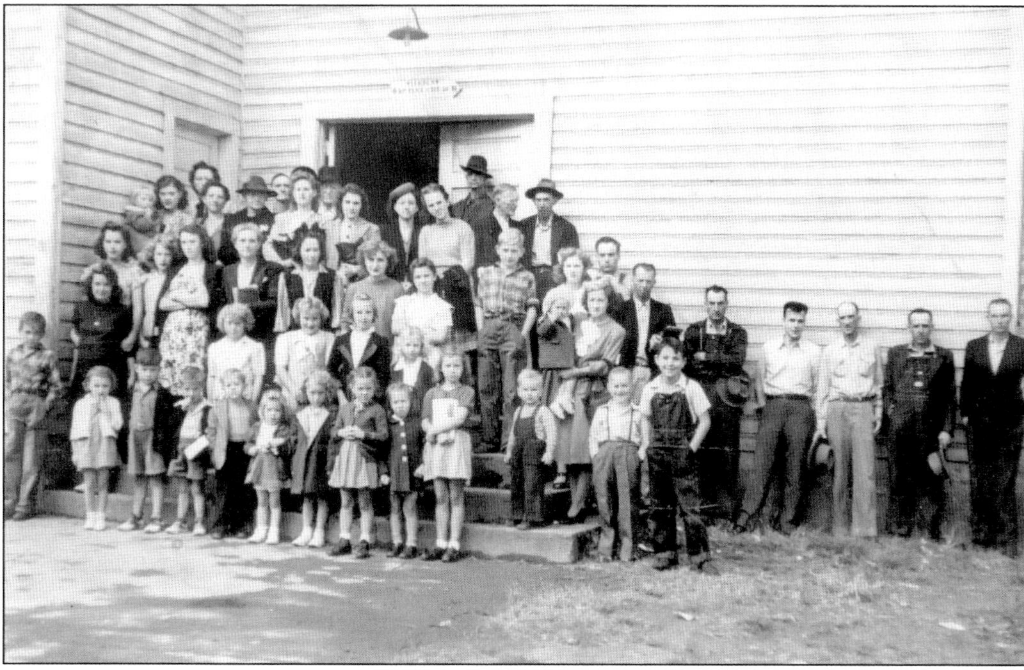

Pictured here is the congregation of Little Milligan Baptist Church. (Photograph courtesy of Juanita Tester Wilson.)

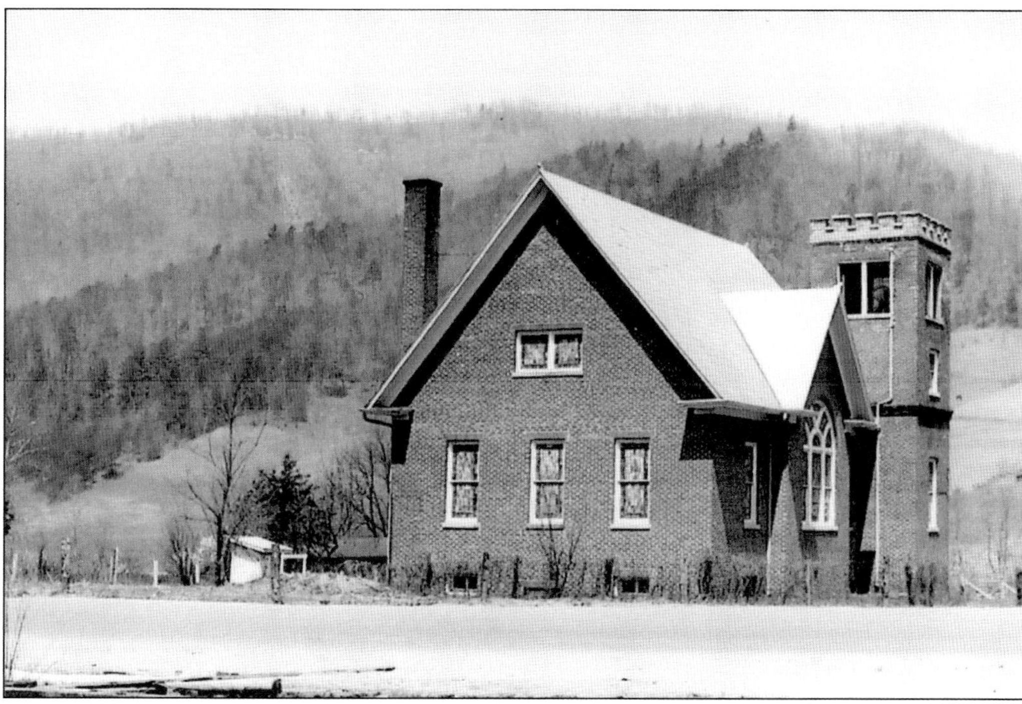

Fish Springs Baptist Church has served the community of Fish Springs since 1914. During the late 1940s, the building was torn down and a new one was built above the flood line of the Watauga Reservoir. (TVA photograph courtesy of Butler Museum.)

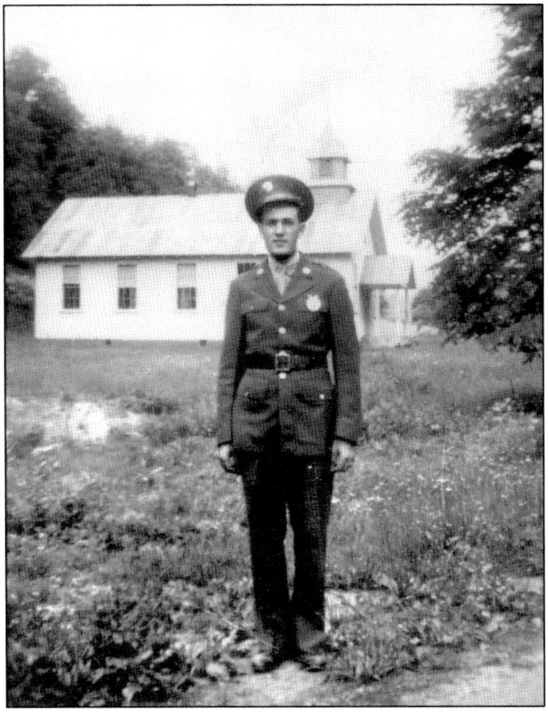

Though the congregation of the Midway Baptist Church endured a serious decline in attendance when families relocated out of the Watauga Reservoir, it still prospers today. In this 1940s photograph, Carl Stanton stands near the church building that served as the church's meeting place for more than 70 years. (Photograph courtesy of Juanita Tester Wilson.)

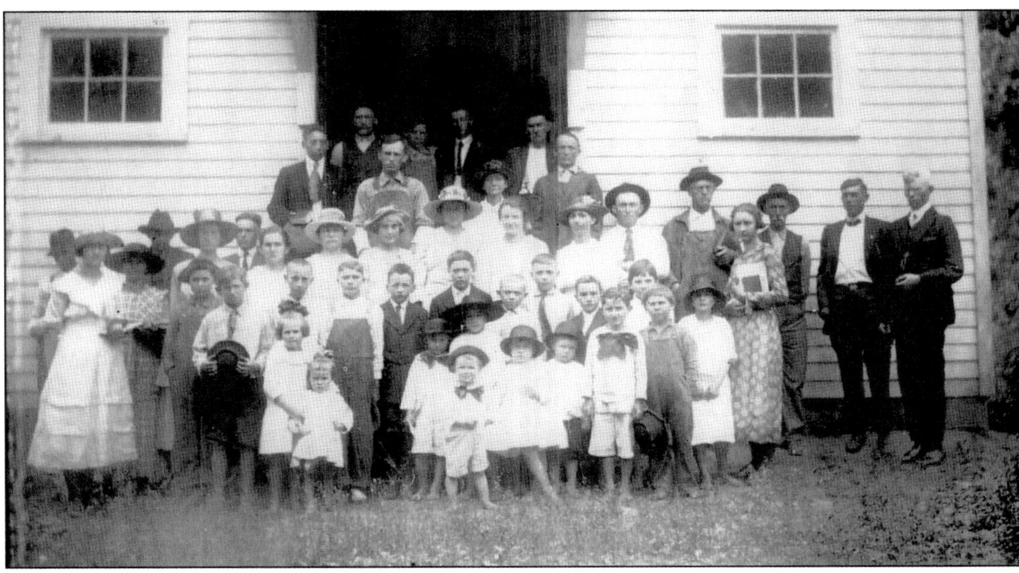

Included in this photograph of the congregation of the Midway Baptist Church are the following: Haskel Arney, Duke Arney, Lena Grindstaff, Myrtle Grindstaff, Berlyn Grindstaff, Novelee Grindstaff, Curtis Forrester, Stanley Forrester, Nub Bradley, Goose Bradley, Burnice Forrester, Ordie Greenwell, Joann Forrester, Ada Grindstaff, Mae Bradley, Neta Bradley, Tina Bradley, Veril Cardwell, Maggie Grindstaff, Bertie Bradley, Wiley Bradley, James Greenwell, Texie Forrester, Samuel Forrester, McKinley Laws, Clint Forrester, Ora Fletcher, Dock Cable, Dan Grindstaff, Stacy Grindstaff, Amanda Forrester, Roscoe Forrester, Ebb Forrester, Gideon Cardwell, Frank Jewett, and Frank Greenwell. The photograph was taken in 1921 in front of Midway School, where the congregation originally met. (Photograph courtesy of Juanita Tester Wilson.)

Five

People and Places

The creeks and rivers, hillsides, and valleys provided a variety of activities for the residents of the Watauga Valley. In the winter, young people could be found building snowmen or walking along the snow-covered paths. In the summer, local swimming holes were a refreshing place to cool off and relax on a hot day. Local businesses such as the Curtis Drugstore and the City Barber Shop provided plenty of opportunity for residents young and old to congregate and catch up on all the local news and gossip. Yet despite the seemingly carefree lifestyle, Butler was never disconnected from the rest of the world, and when the call to duty was issued, young men who responded were seen in town on furlough and preparing to go overseas. Butler residents joined and participated in all branches of service in the Civil War, Spanish-American War, World War I, and World War II. The myriad of family portraits and social gatherings in Butler illustrates a close-knit sense of community that existed in the Watauga Valley.

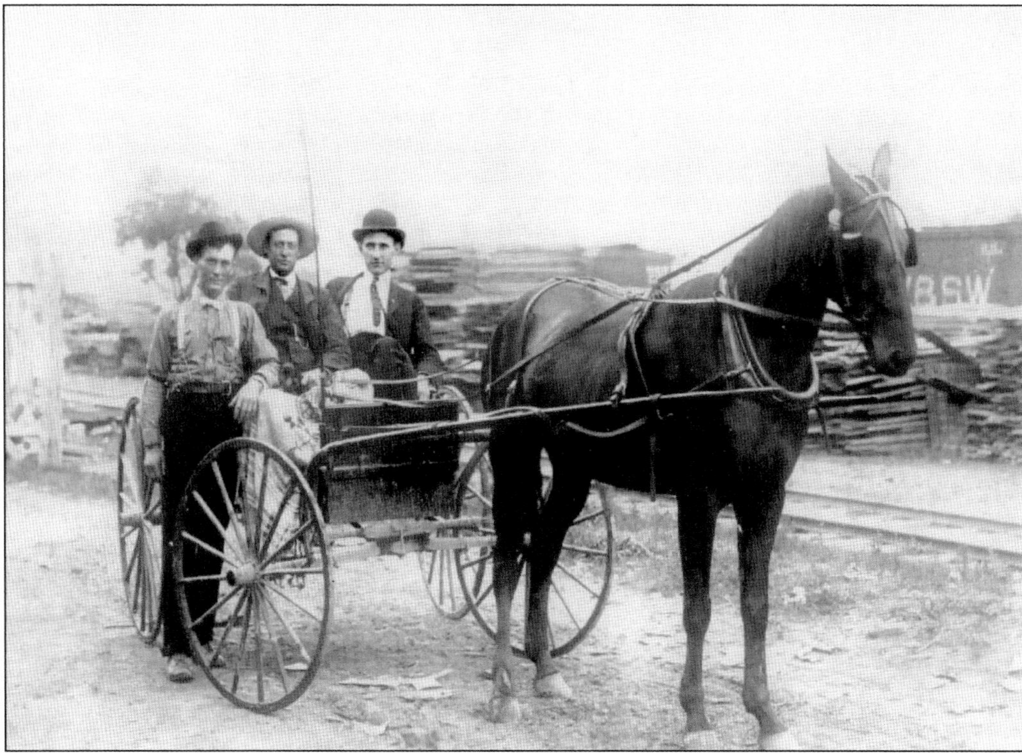

Following World War I, many of the young men of the Watauga Valley attempted to satisfy their "wanderlust" by traveling west. Pictured here from left to right are Stan Atwood, Troy Atwood, and Smith Dugger at the Butler Depot. (Photograph courtesy of Butler Museum.)

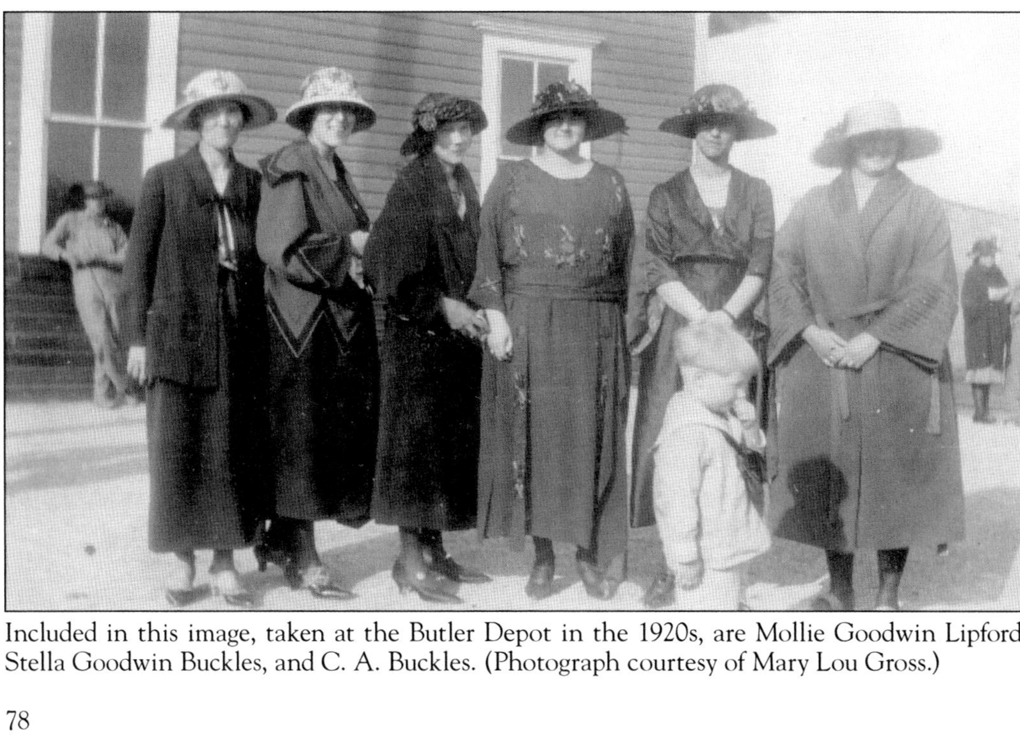

Included in this image, taken at the Butler Depot in the 1920s, are Mollie Goodwin Lipford, Stella Goodwin Buckles, and C. A. Buckles. (Photograph courtesy of Mary Lou Gross.)

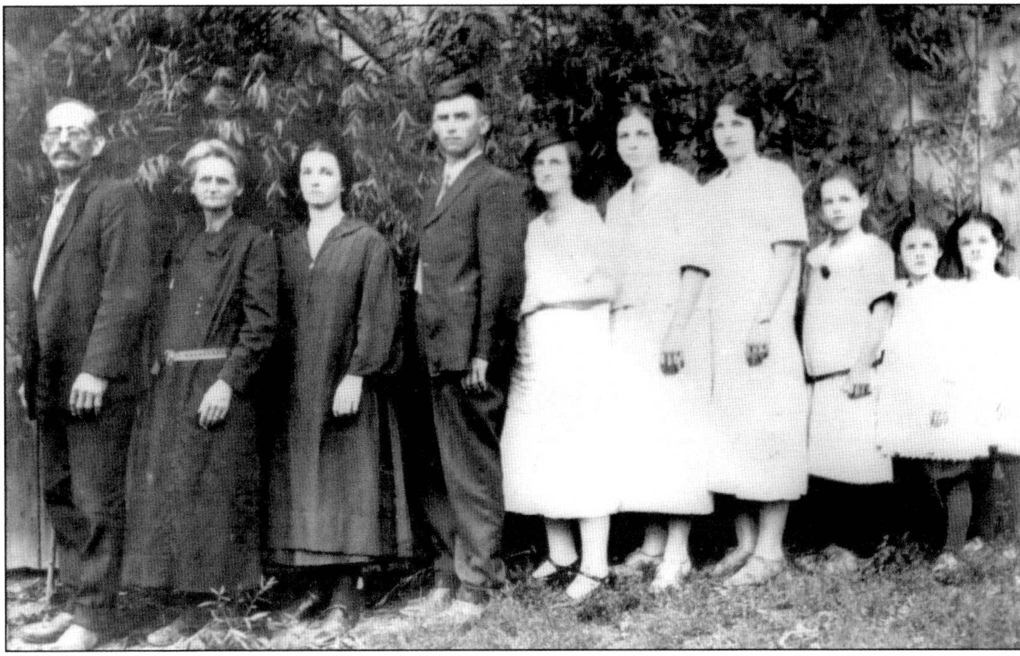

William Cable (far left) and Ella Anderson Cable (second from left) of Sugar Grove are pictured here with their children, from left to right, Nettie Cable Main, Dewey Cable, Annie Cable Conway, Ordie Cable Potter, Verdie Cable Potter, Delores Cable Tester, Alma Cable Ward, and Alta Cable Triplett. (Photograph courtesy of Juanita Tester Wilson.)

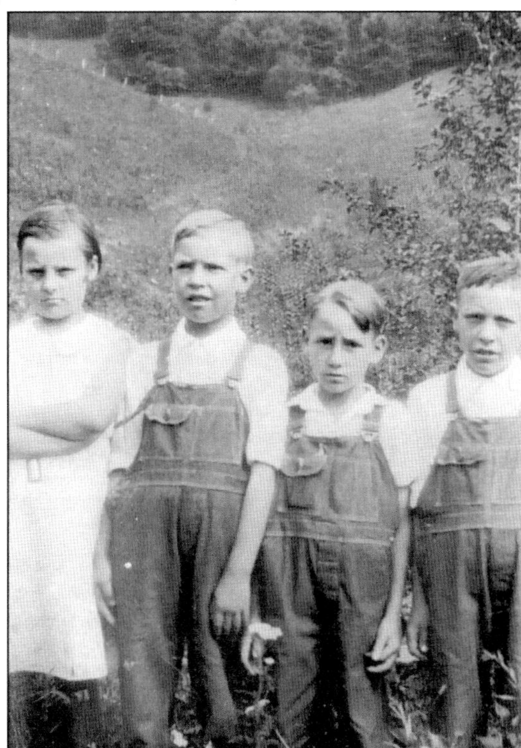

These children from the Sugar Grove community are, from left to right, Lucille Conway, Loyd Gregg, Carter Isaacs, and Lynell Gregg. (Photograph courtesy of Juanita Tester Wilson.)

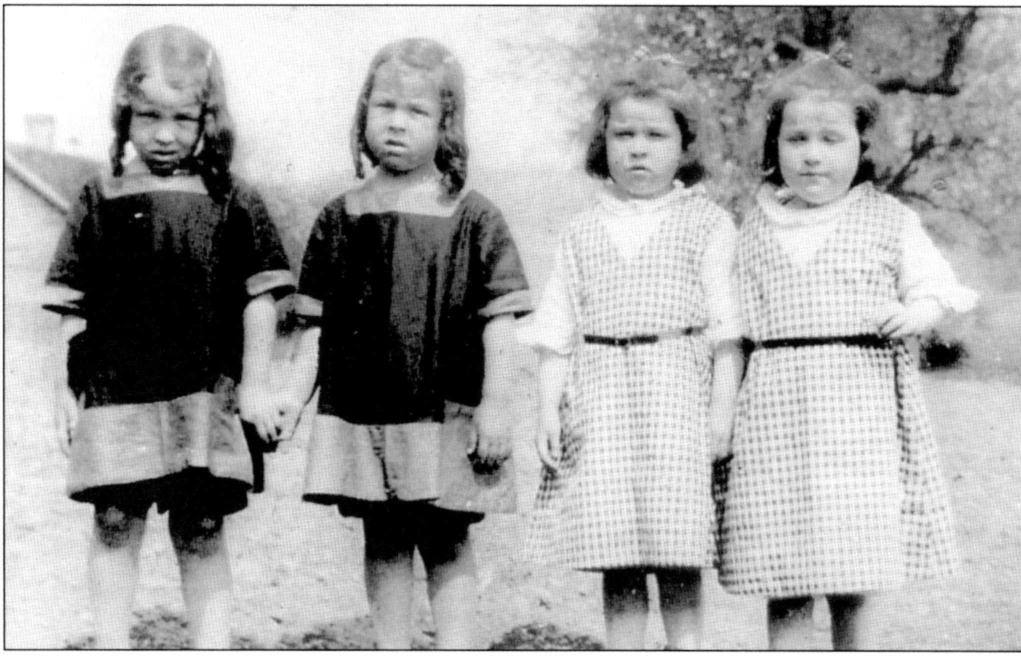

Pictured here, from left to right, are Dorothy and Doris DeVault and Alma and Alta Cable. The DeVault twins were the daughters of Robert Martin DeVault Sr., the pastor of Sugar Grove Baptist Church. The Cable twins were the daughters of William and Ella Cable. (Photograph courtesy of Juanita Tester Wilson.)

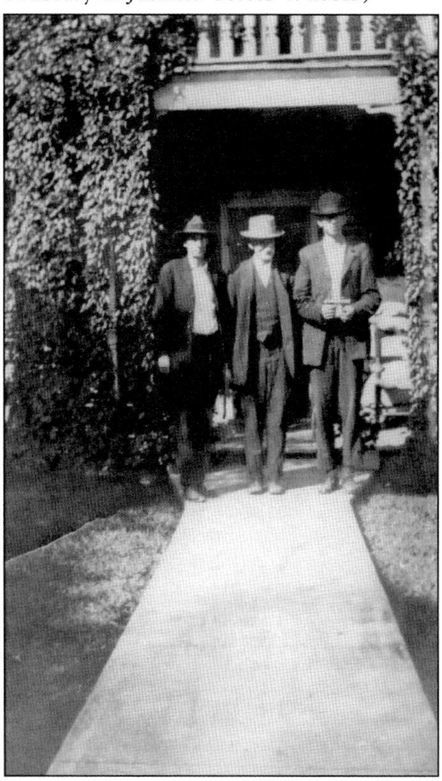

Glenn Walker (left), Allen Carriger (center), and George T. Walker (right) stand in front of the Carriger family home. (Photograph courtesy of George Walker.)

C. A. Buckles (front), Stella Goodwin Buckles (left), John Franklin Goodwin (center), and Mary Lucinda Shoun Goodwin (right) are pictured in this 1920s photograph. (Photograph courtesy of Mary Lou Gross.)

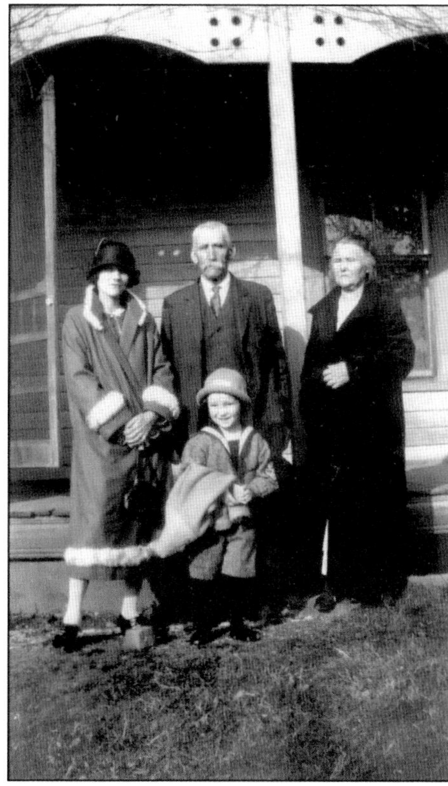

Pictured here from left to right are Pauline Lipford Peters, Austin Goodwin, Lucy Hughes, Jean Hughes, and Joyce Hughes. (Photograph courtesy of Mary Lou Gross.)

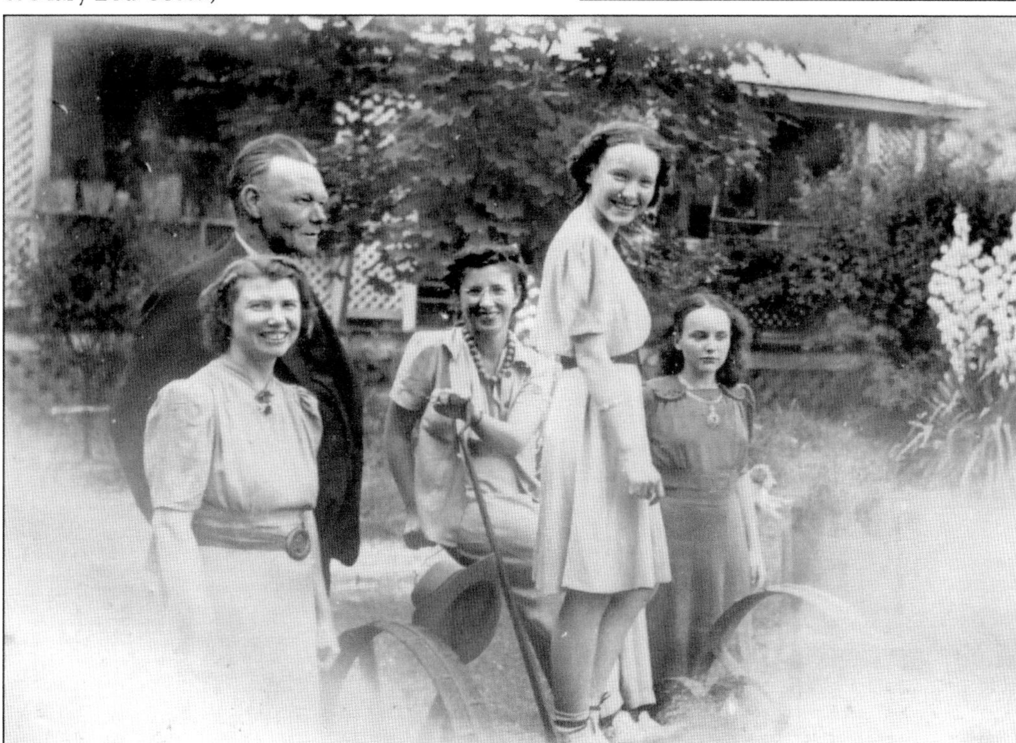

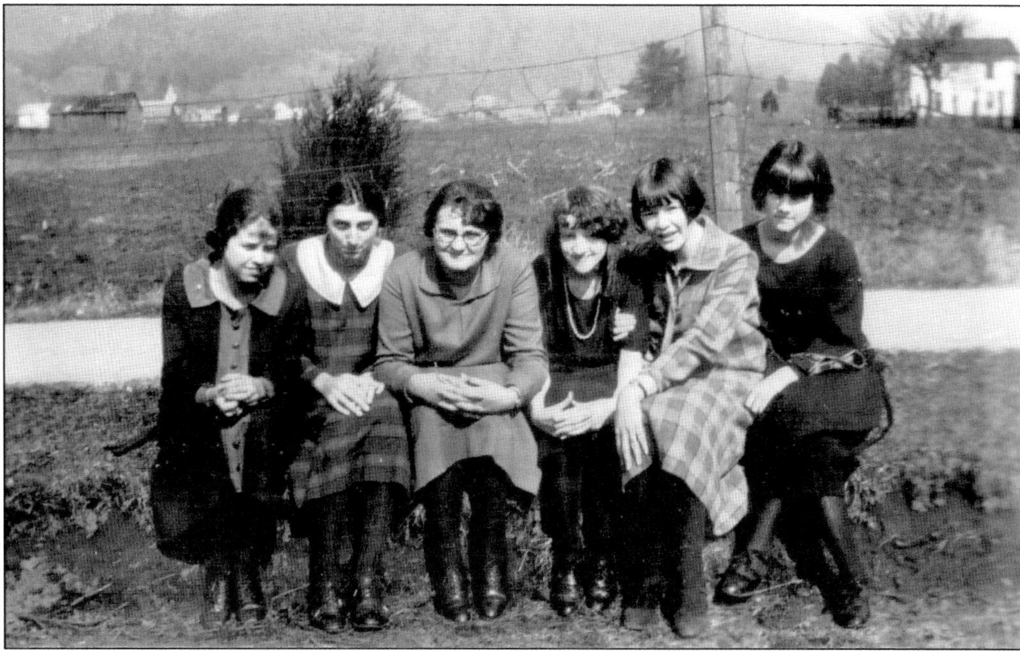

This photograph, found in Mary Goodwin's memory book, is of her graduation from Watauga Academy in 1925. Pictured from left to right are Ethel Dugger, Mary Goodwin, Hildred Wagner, Rheta Williams, Ruth Goodwin, and Aileen King. (Photograph courtesy of Butler Museum.)

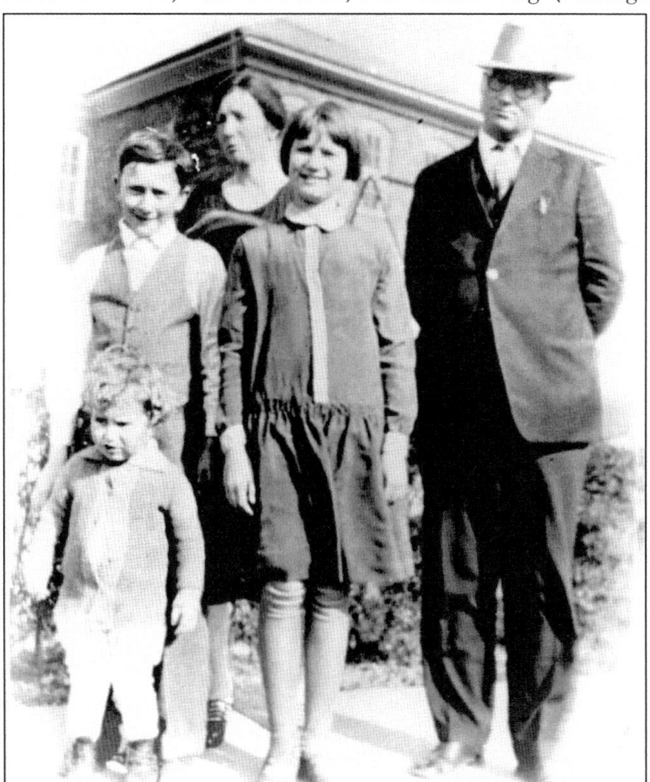

This photograph, taken in 1925, is of Ira Q. Harris and his family. Pictured from left to right are (first row) Albert Harris; (second row) James C. Harris and Margaret Harris Robinson; (third row) Margaret Smythe Harris and Ira Q. Harris. The Harris family lived on Church Street across from Butler City School. (Photograph courtesy of James D. Robinson.)

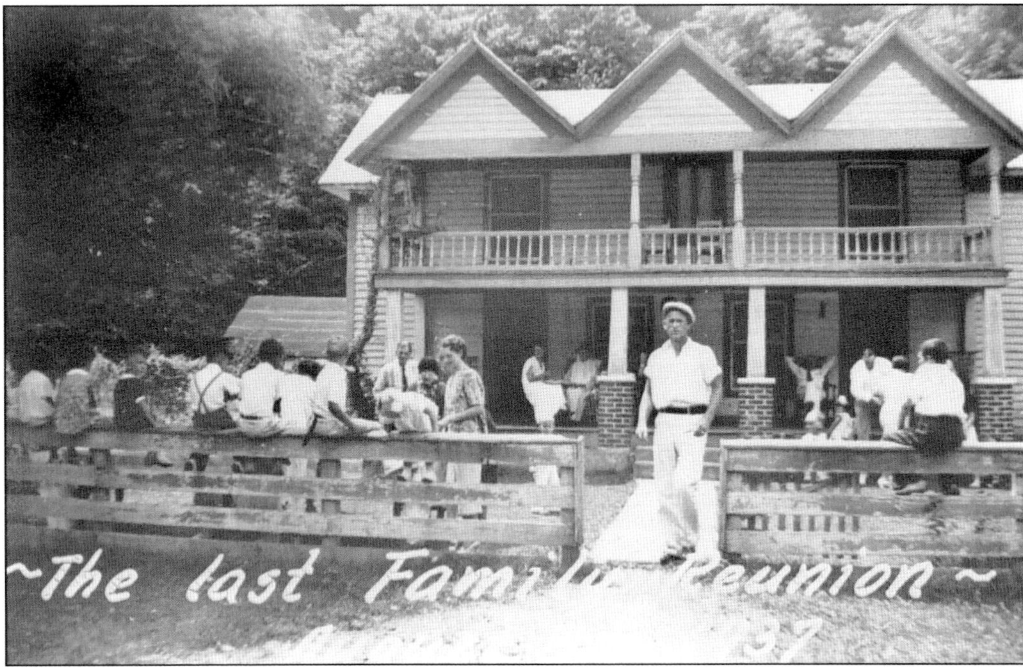

The last family reunion of the L. L. McQueen family, held on August 8, 1937, is shown here. Shortly after the reunion, L. L. McQueen died on October 3, 1937. (Photograph courtesy of James D. Robinson.)

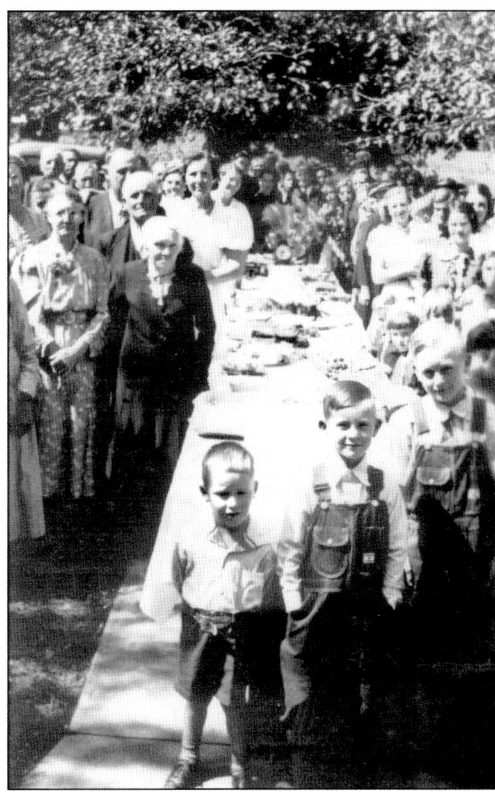

In this picture, "Aunt Jane" Hazelwood celebrates her 75th birthday at the home of her daughter, Mrs. Carrie Jones, on Main Street. (Photograph courtesy of Butler Museum.)

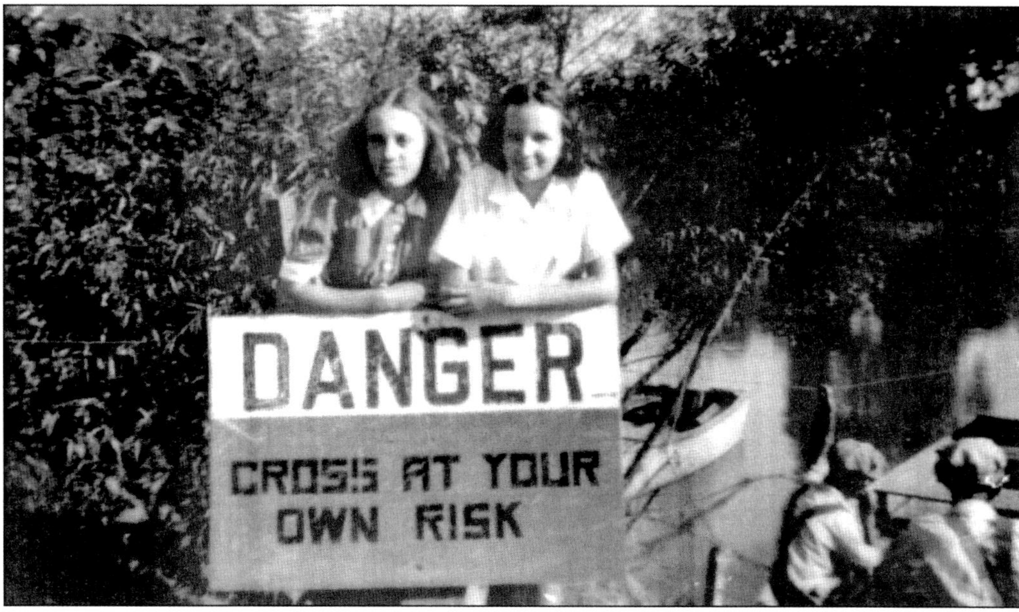

During the 1930s, the Butler Bridge had to be replaced. Helen Stout (left) and Jo Goodwin (right) pose near the Watauga River as workmen ferry people across. (Photograph courtesy of Butler Museum.)

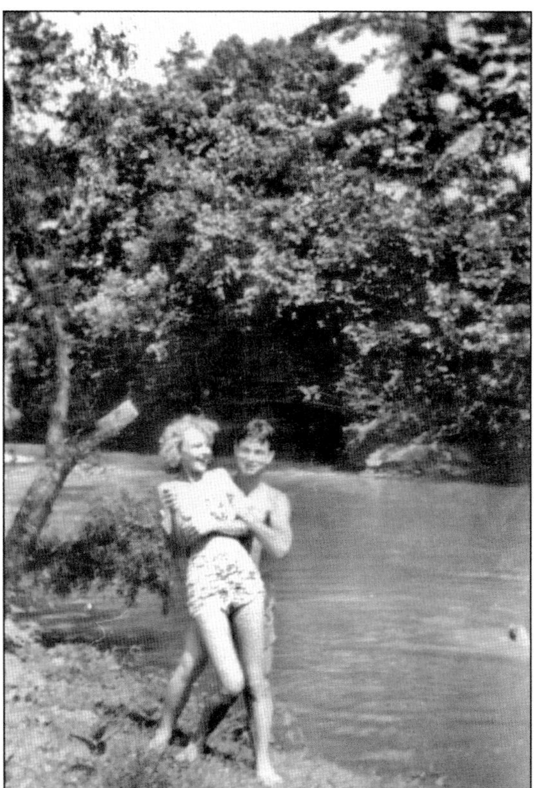

The Curtis swimming hole was located on Roan Creek across from the Curtis house. (Photograph courtesy of Butler Museum.)

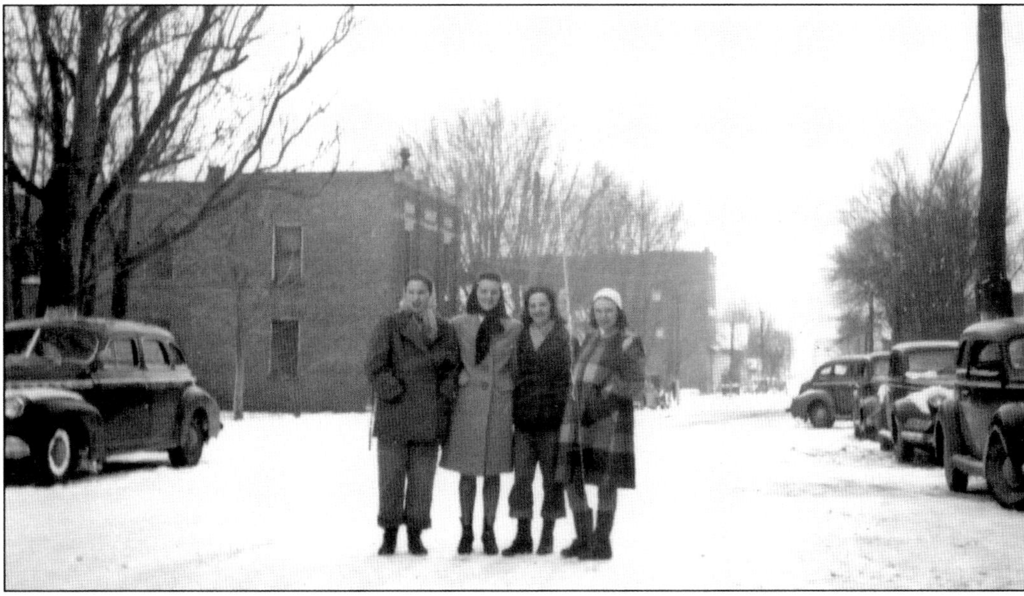

From left to right, Florence Tucker, Ester Arney, Edna Griffitts, and Selma "Babe" Curtis pose in the snow on Main Street in the 1940s. (Photograph courtesy of Butler Museum.)

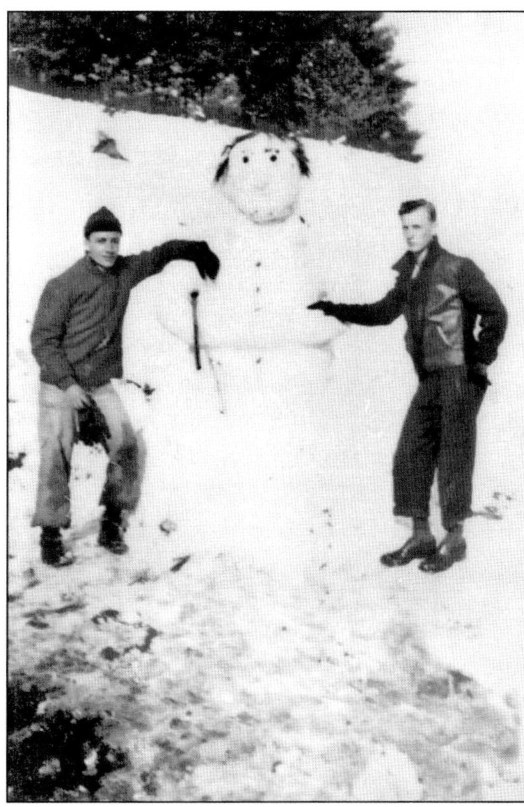

Pictured here are Bill Trivette (left) and Frankie Ramsey, graduates of Watauga Academy. (Photograph courtesy of Butler Museum.)

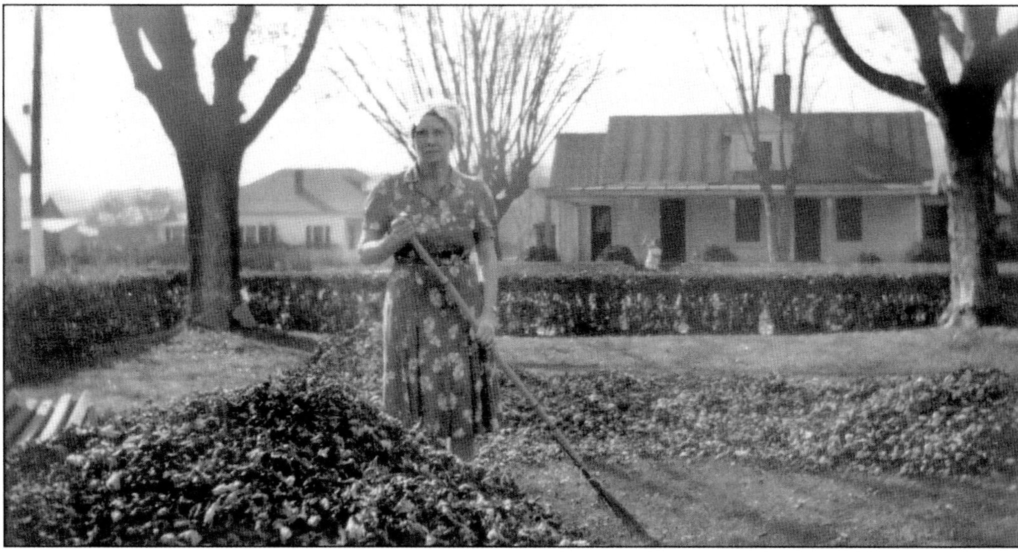

This photograph of Rose Trivette was taken on Spring Street. The buildings in the background are, from left to right, the Methodist church, Lee Cable's house, and Dr. James Robinson's house. (Photograph courtesy of Butler Museum.)

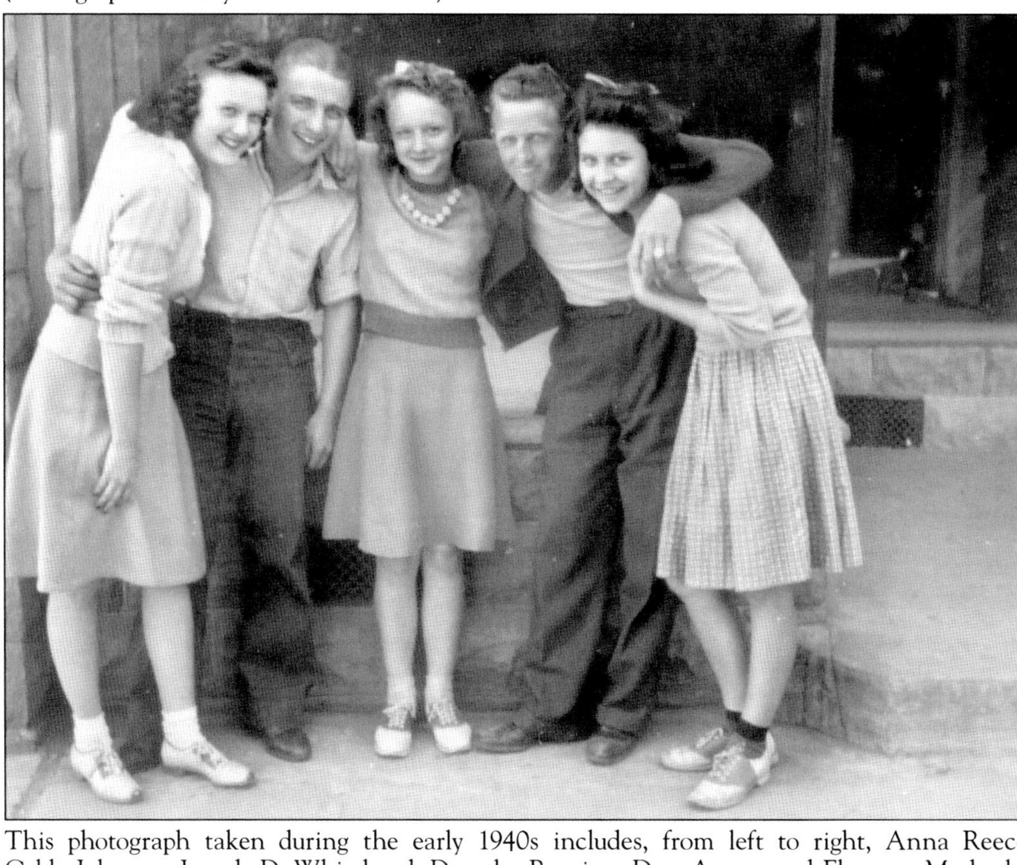

This photograph taken during the early 1940s includes, from left to right, Anna Reece Cable Johnson, Joseph D. Whitehead, Dorothy Romine, Don Arney, and Florence Matherly. (Photograph courtesy of Butler Museum.)

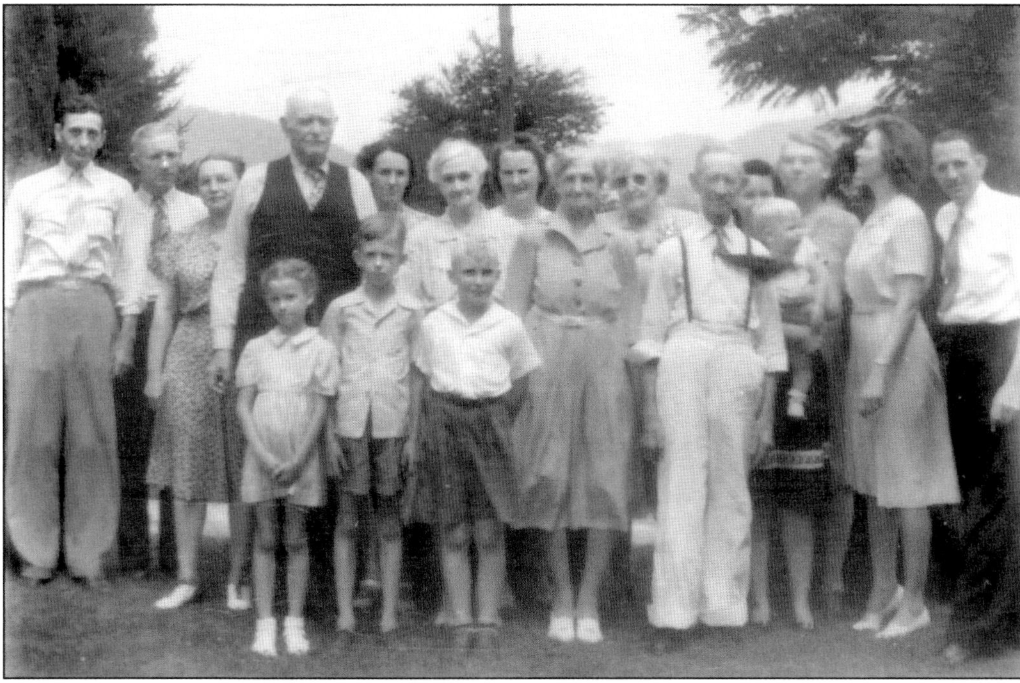

Included in this photograph, taken at the Joe Pierce house, are Vasco Howell, A. C. Ward, Cora Lee Ward, Lee Goodwin, Mrs. Lee Goodwin, Cora Pierce, Joe Pierce, Mamie Fine, and Ruth Howell. (Photograph courtesy of Butler Museum.)

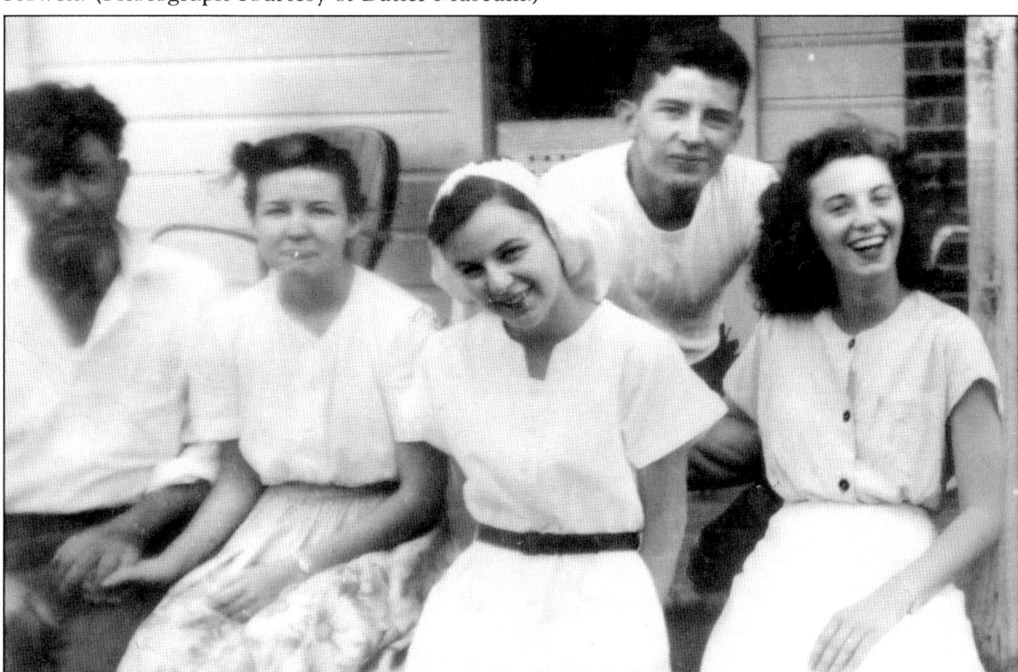

Pictured here, from left to right, are Clay Dugger, Anna Dugger, Wilma Dugger, Charles Dugger, and Mary Ellen Walker. Anna, Charles, and Mary Ellen were graduates of the 1947 class of Watauga Academy. (Photograph courtesy of Mary Walker Ward.)

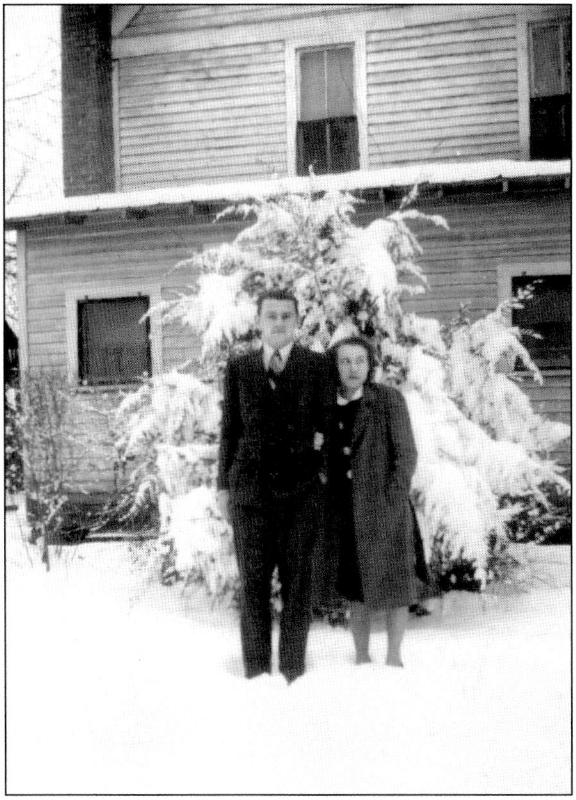

Pictured here are B. A. Lipford Jr. (left) and Leondas Farthing Lipford. The Lipford family lived in the Dry Hill community. (Photograph courtesy of Patsy Tadlock and Ermine Arwood.)

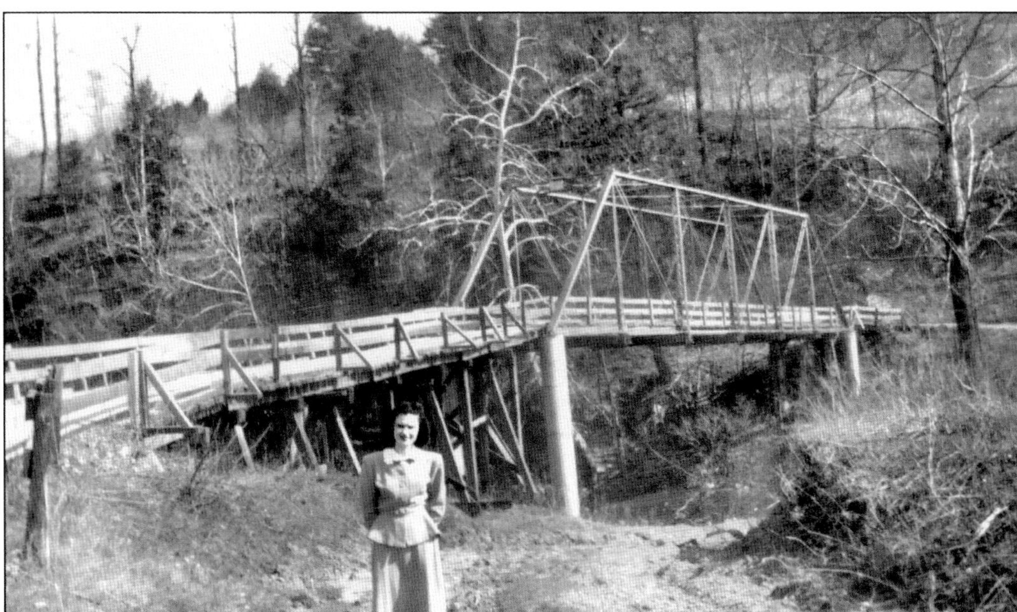

Lucille McQueen stands near the Roan Creek Bridge in this picture, which was taken in 1948. (Photograph courtesy of Butler Museum.)

Spanish-American War veteran David Day poses in front of the Whiting Lumber Company in 1919. (Photograph courtesy of Mary Walker Ward.)

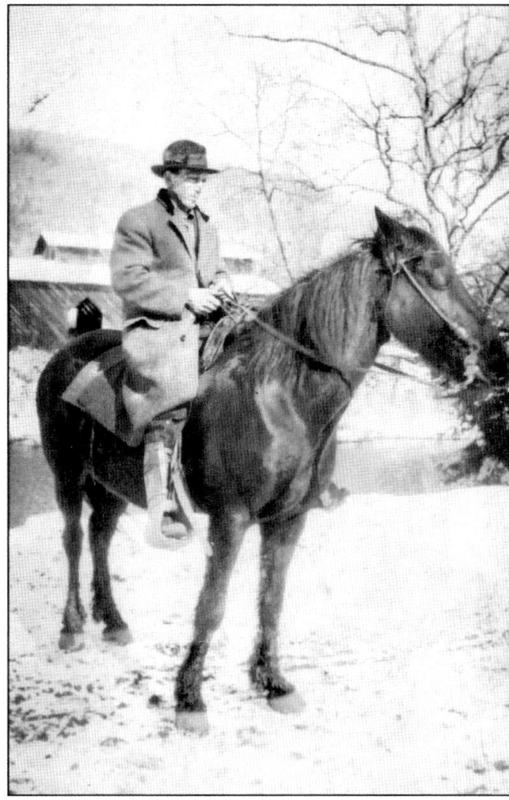

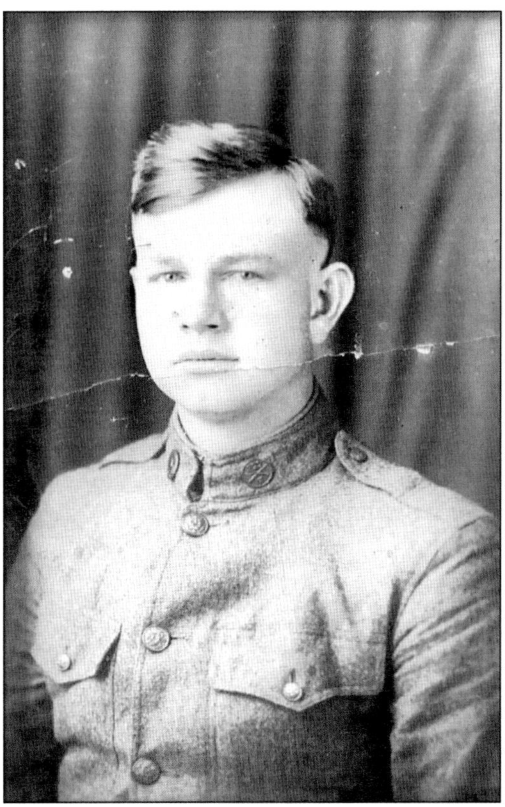

Frank Greenwell served in the U.S. Army during World War I. (Photograph courtesy of Juanita Tester Wilson.)

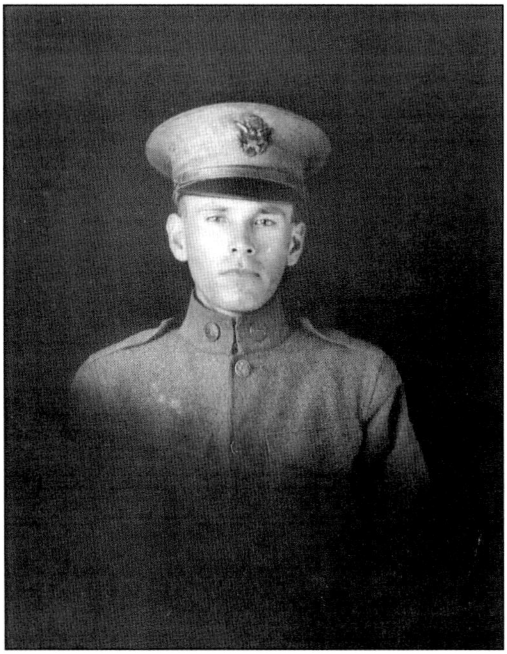

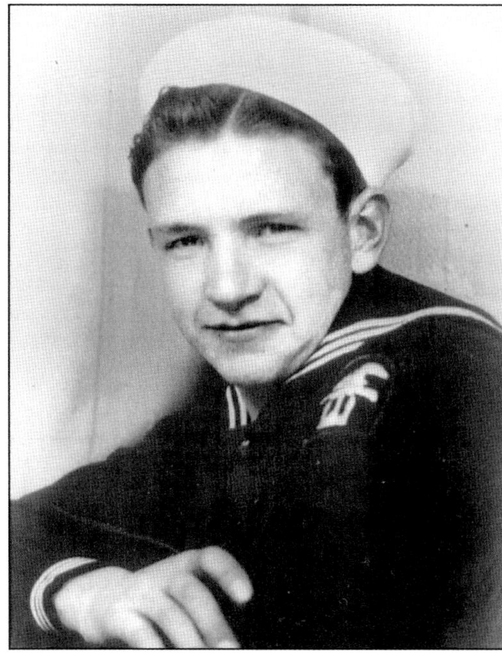

(Above, left) William Elbert Forrester served in World War I. Afterwards he returned to the Midway community, married Texie Greenwell, and became a veterinarian. (Photograph courtesy of Juanita Tester Wilson.)

(Above, right) Dexter Forrester, son of William Elbert and Texie Forrester, served in the navy during World War II. (Photograph courtesy of Juanita Tester Wilson.)

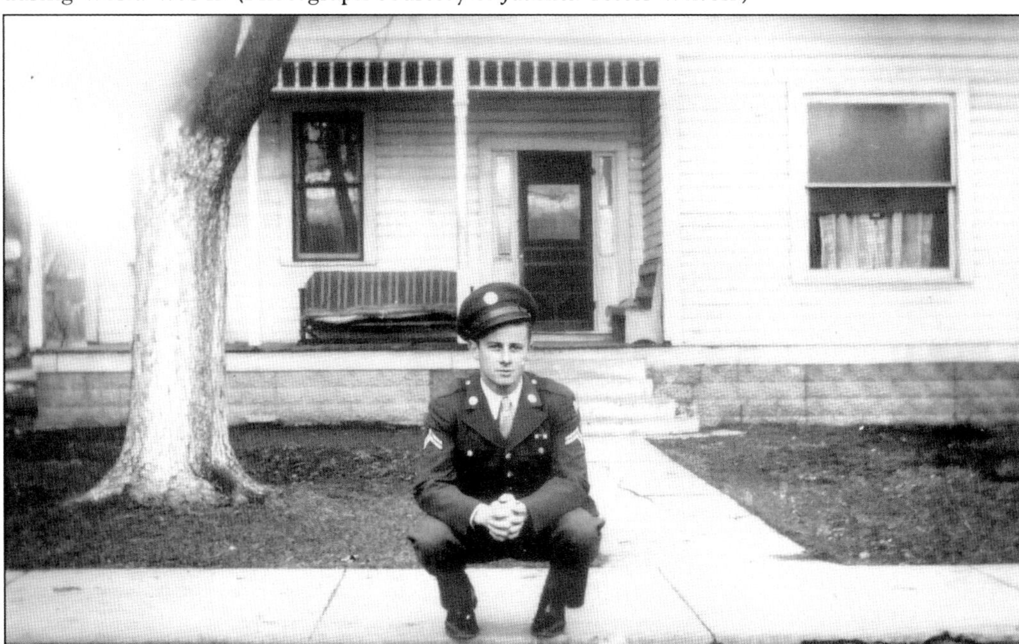

Cpl. Ray Trivette poses in front of the Trivette home on Spring Street in this 1940s picture. (Photograph courtesy of Butler Museum.)

Haynes Greenwell visits his hometown on leave from the navy. (Photograph courtesy of Butler Museum.)

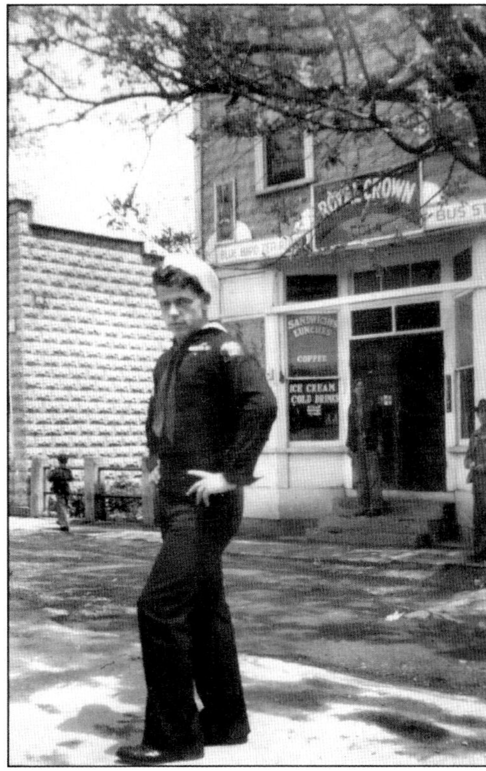

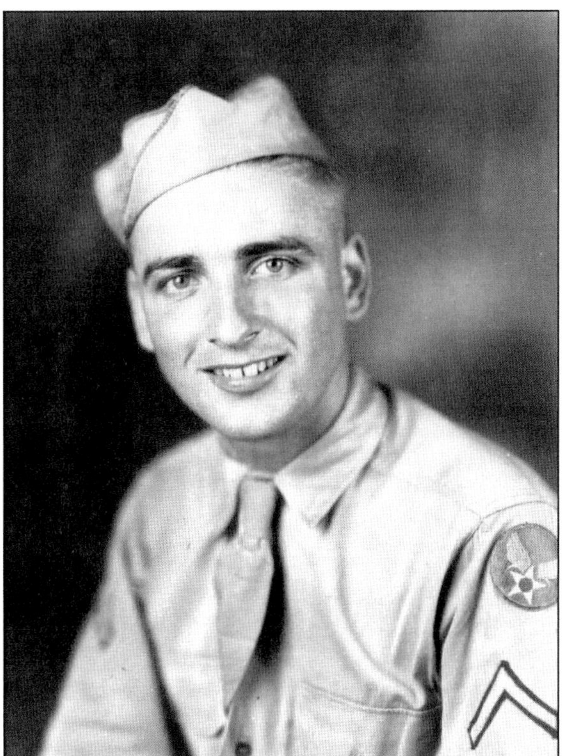

J. D. Whitehead served in the Army Air Corps during World War II. (Photograph courtesy of Butler Museum.)

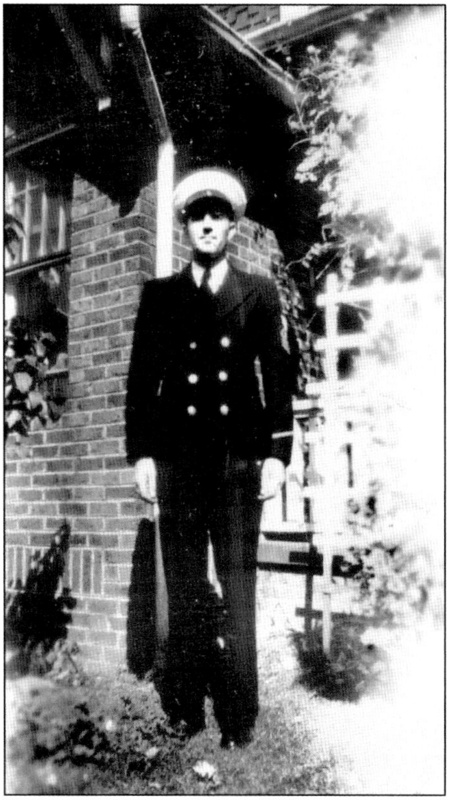

CPO Ron Johnson served in the navy during World War II. (Photograph courtesy of Butler Museum.)

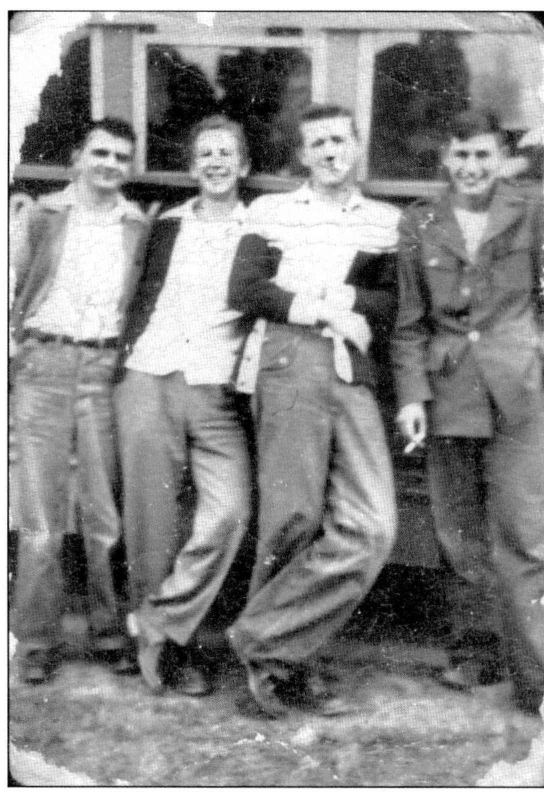

The young men in this photograph, taken in the early 1950s, are, from left to right, Ed Grindstaff, Herman Tester, Jack Cable, and Vestal Cowan. Jack and Ed attended Watauga Academy. Vestal was killed in the Korean War. (Photograph courtesy of Juanita Tester Wilson.)

Six

Floods and the Tennessee Valley Authority

When Franklin Delano Roosevelt came into office in 1933, the nation was expecting dramatic changes. President Roosevelt's New Deal was designed to help America through the worst depression the nation had ever experienced. One of the proposed programs under the New Deal was the Tennessee Valley Authority (TVA). TVA was designed to be a series of dams and power plants along the rivers of the Tennessee Valley. The TVA had two primary goals: the first was to provide affordable electrical power to the Tennessee Valley, which follows the path of the Tennessee River and its tributaries through Kentucky, Virginia, North Carolina, Tennessee, Georgia, Alabama, and Mississippi; the second was to provide flood control for the Tennessee Valley. Even though the Watauga River had provided power to the region through the use of waterwheels, the river was truly untamed and often ravaged the Watauga Valley suddenly and unexpectedly with devastating floods. The Watauga Dam project brought an end to the devastation of periodic floods in the areas. The TVA became one of the most successful programs of the New Deal, and it remains with us today.

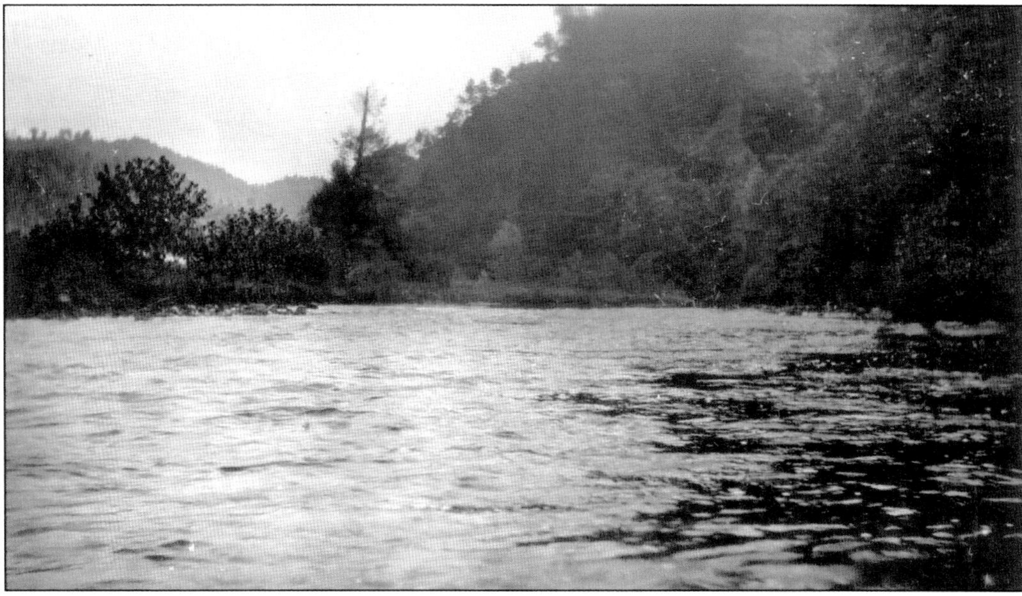

Many homes and lives have been lost through the years due to the Watauga River. Notorious floods during the 20th century occurred in 1901, 1924, and 1940. (Photograph courtesy of Butler Museum.)

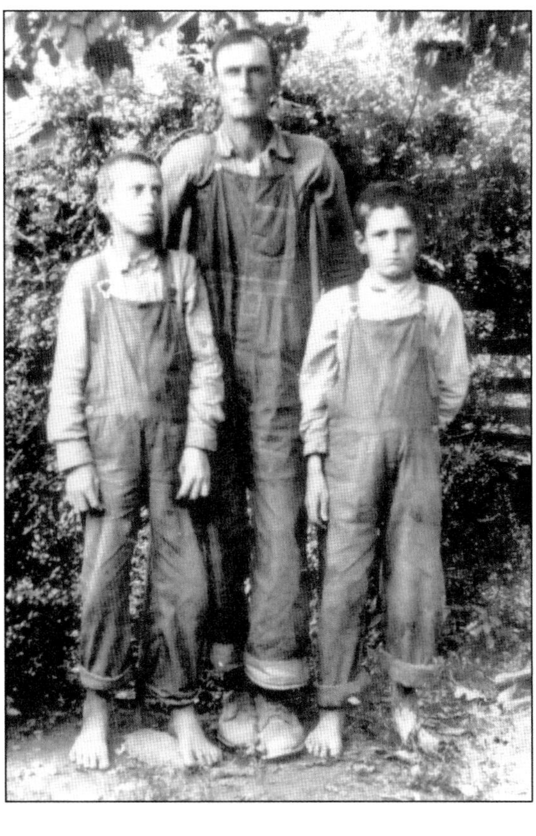

Cecil Lewis is pictured here with his two sons, Walter (left) and Charles. Twelve people died as a result of the flood of 1924, including the other six members of Cecil's family. (Photograph courtesy of Butler Museum.)

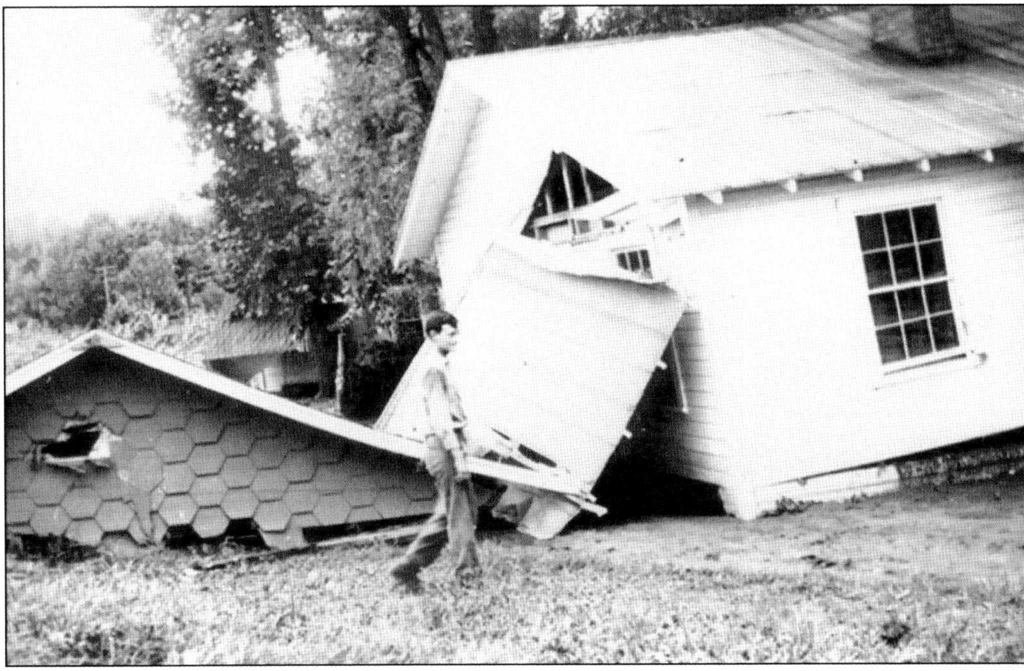

In August 1940, the Watauga River swelled once again beyond its banks and decimated the Watauga Valley. Pictured here is one of the many houses that were destroyed as a result of the flood of 1940. (Photograph courtesy of Butler Museum.)

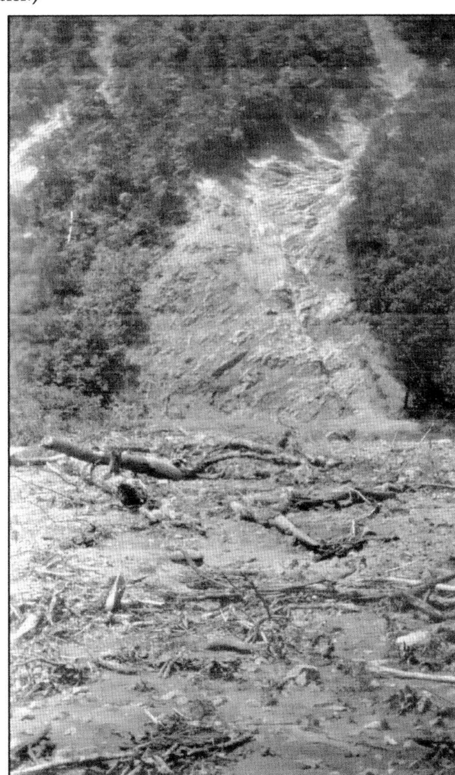

The flood waters were so powerful during the 1940 flood that they uprooted trees and cut paths down the hillsides into the valleys below. (Photograph courtesy of Butler Museum.)

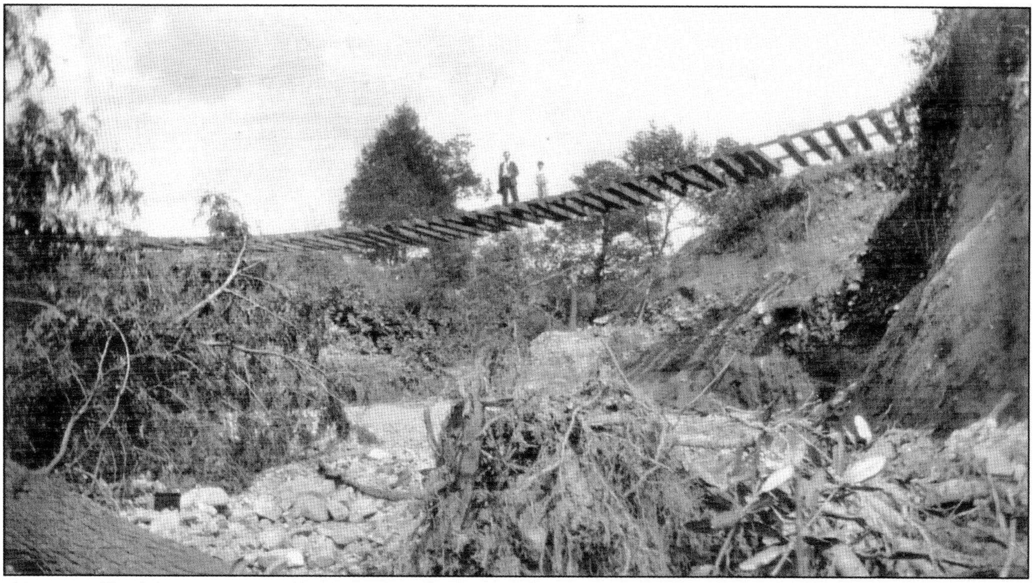

Almost all of the train trestles were washed away by the flood waters and debris during the 1940 flood. (Photograph courtesy of Butler Museum.)

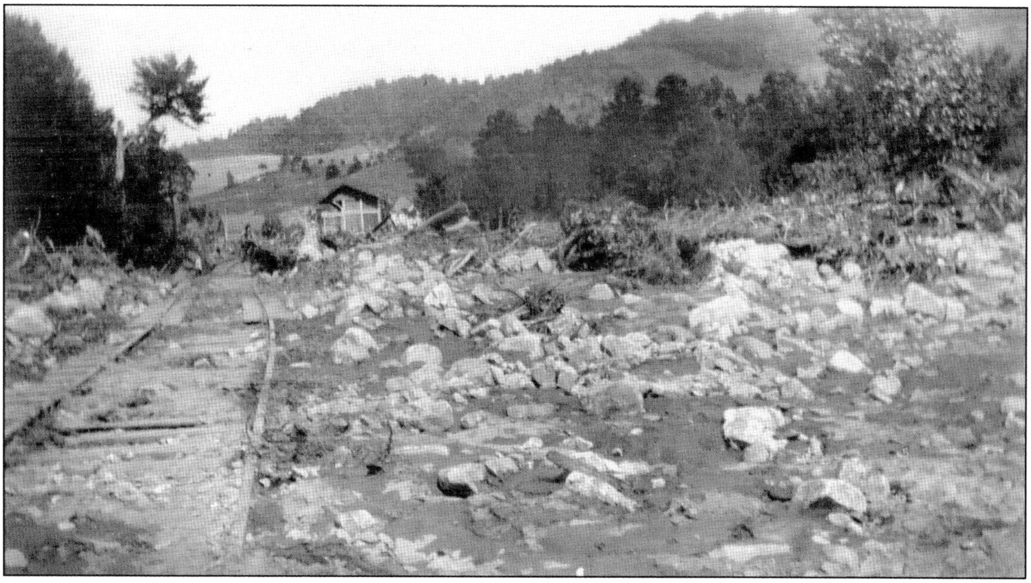

As a result of the devastation of the 1940 flood, Southern Railroad's service in the Watauga Valley was completely halted. Pictured here is Gouge Station, one of many stations that were rendered inoperable. (Photograph courtesy of Butler Museum.)

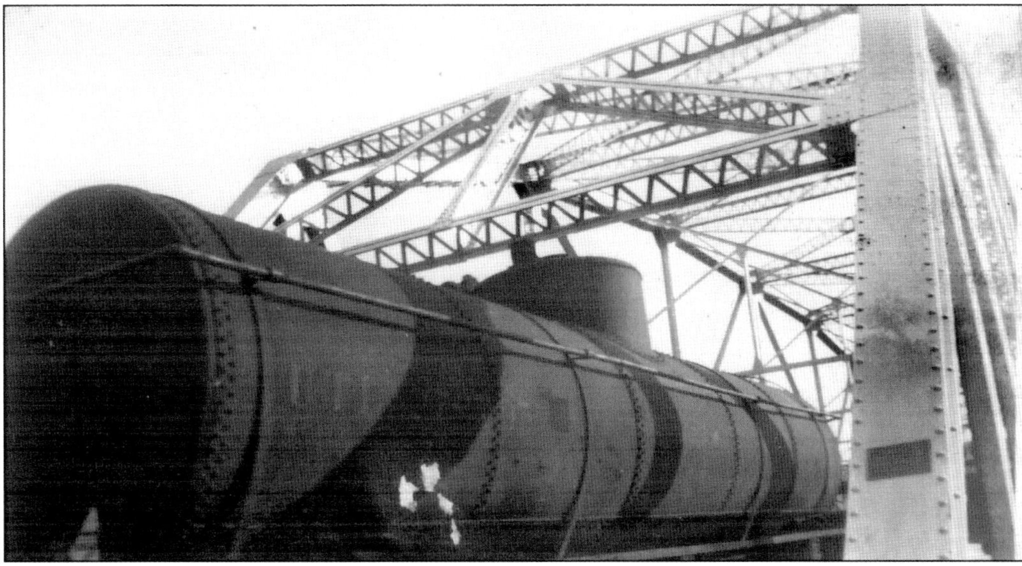

Due to the devastation of the flood, railcars had to be hauled away from Butler using tractor-trailers. This photograph depicts a tanker being transported over the Butler Bridge. (Photograph courtesy of Butler Museum.)

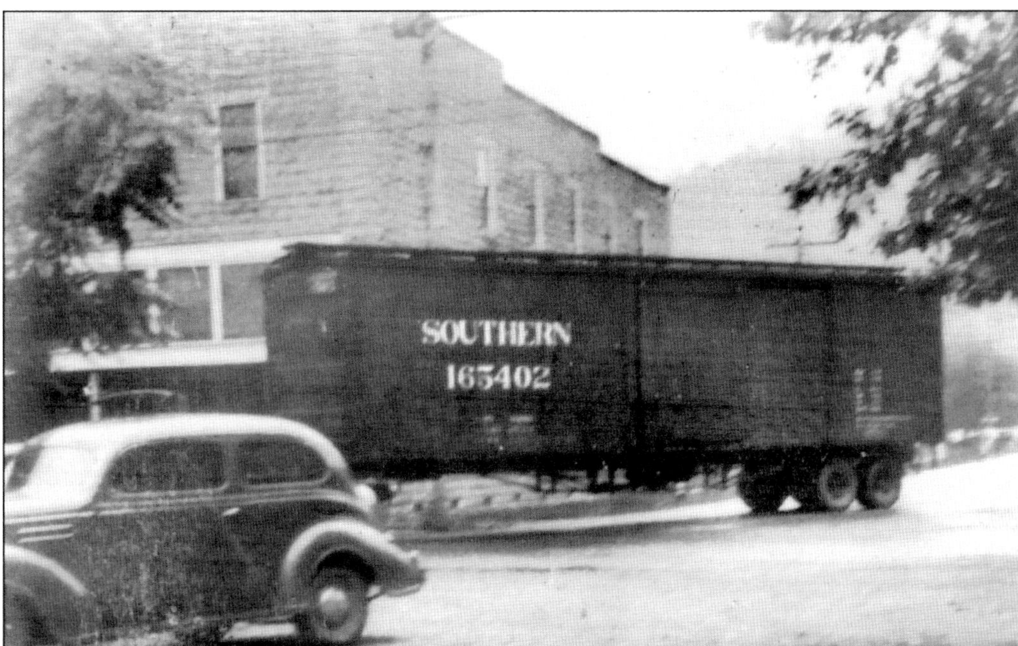

In this photograph, a railcar pulled by a tractor-trailer turns from Spring Street onto Main Street on its way out of Butler. (Photograph courtesy of Butler Museum.)

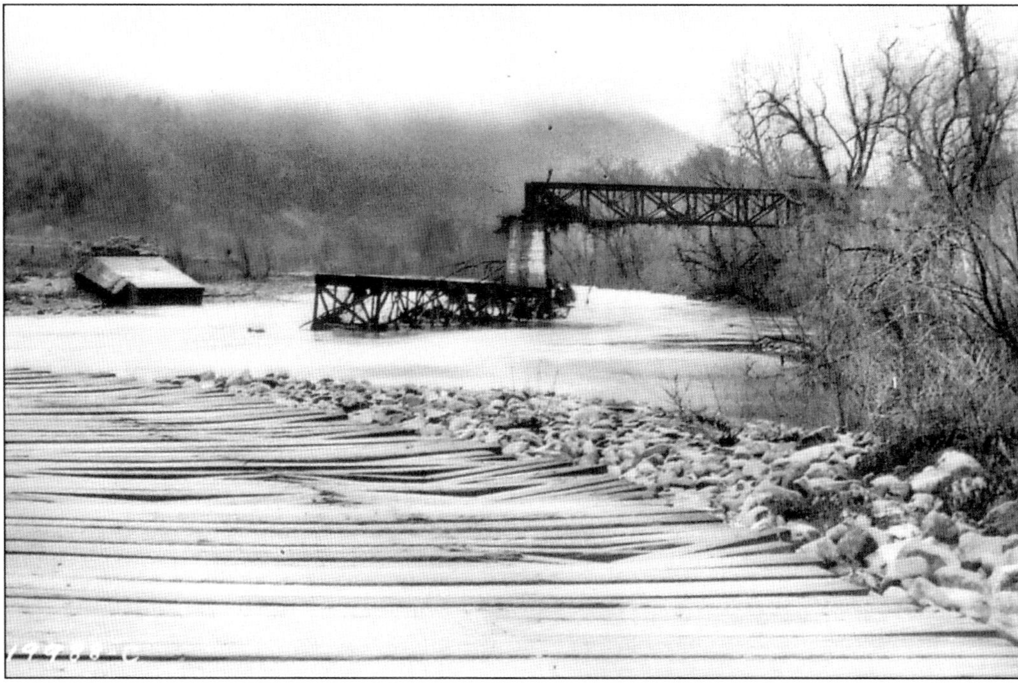

Following the 1940 flood, Southern Railroad chose not to invest time or money restoring the lines through the Watauga Valley. This abandoned railroad bridge was located at Carden's Bluff. (TVA photograph courtesy of Butler Museum.)

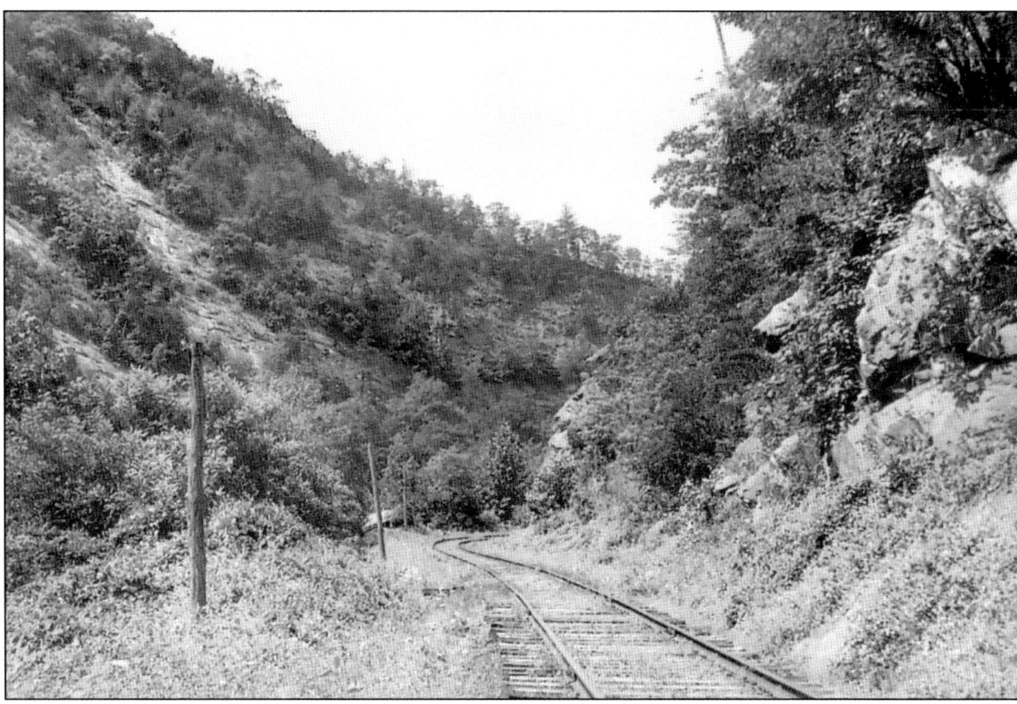

Pictured here are abandoned tracks of the Southern Railroad at Carden's Bluff—the future site of the Watauga Dam. (TVA photograph courtesy of Butler Museum.)

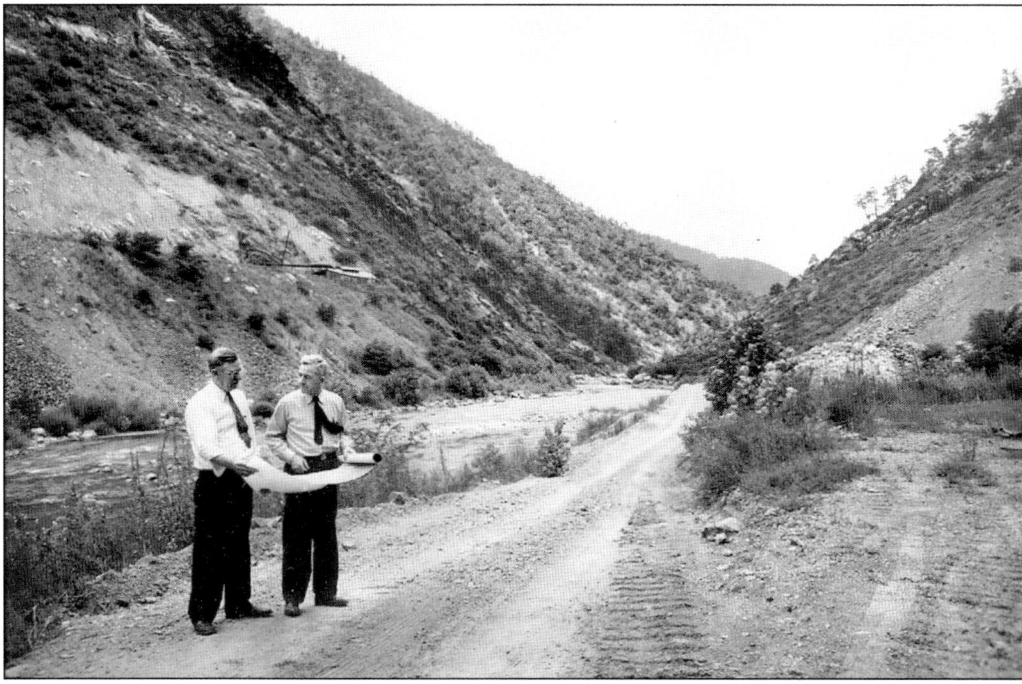

Tennessee Valley Authority engineers are shown here going over the plans for the dam, which would span Carden's Bluff. (TVA photograph courtesy of Butler Museum.)

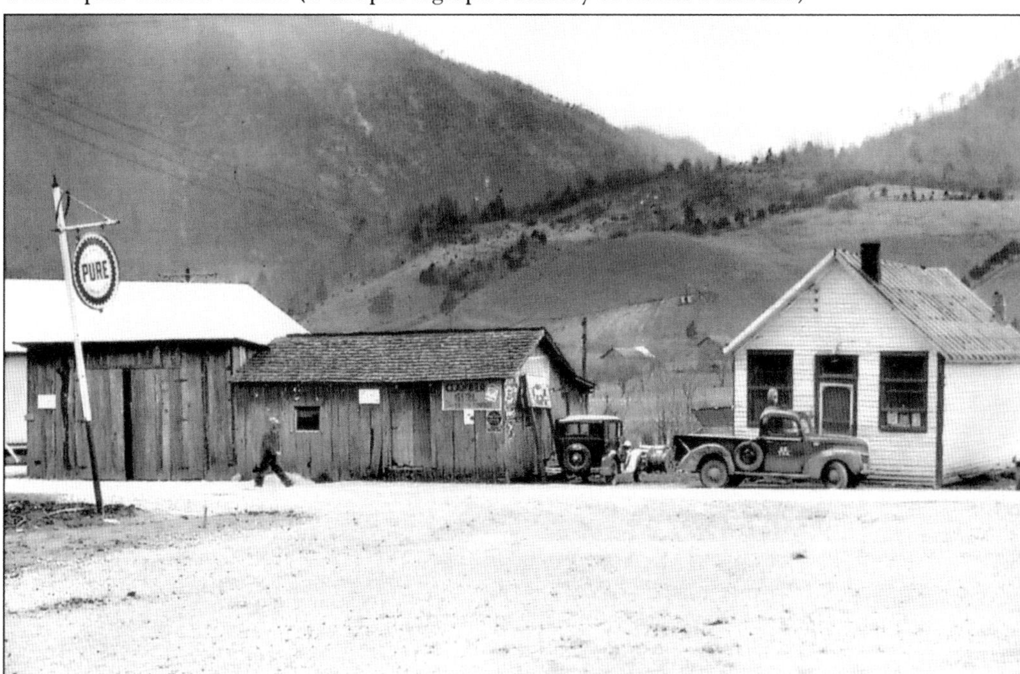

D. B. Pierce's garage and store was located at Carden's Bluff between Butler and Hampton, Tennessee, on Highway 67. Note the Tennessee Valley Authority truck parked in front. Representatives from TVA canvassed the Watauga Valley and recorded residents' opinions about relocation efforts. (TVA photograph courtesy of Butler Museum.)

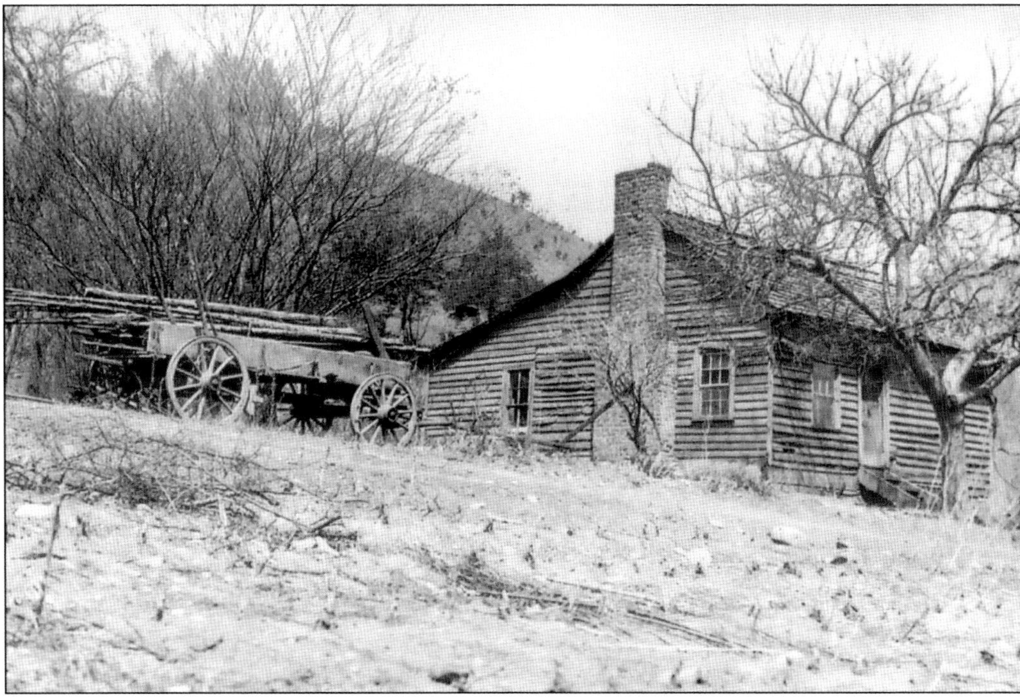

The home of J. F. Lewis was located on Mountain Road in Carden's Bluff. (TVA photograph courtesy of Butler Museum.)

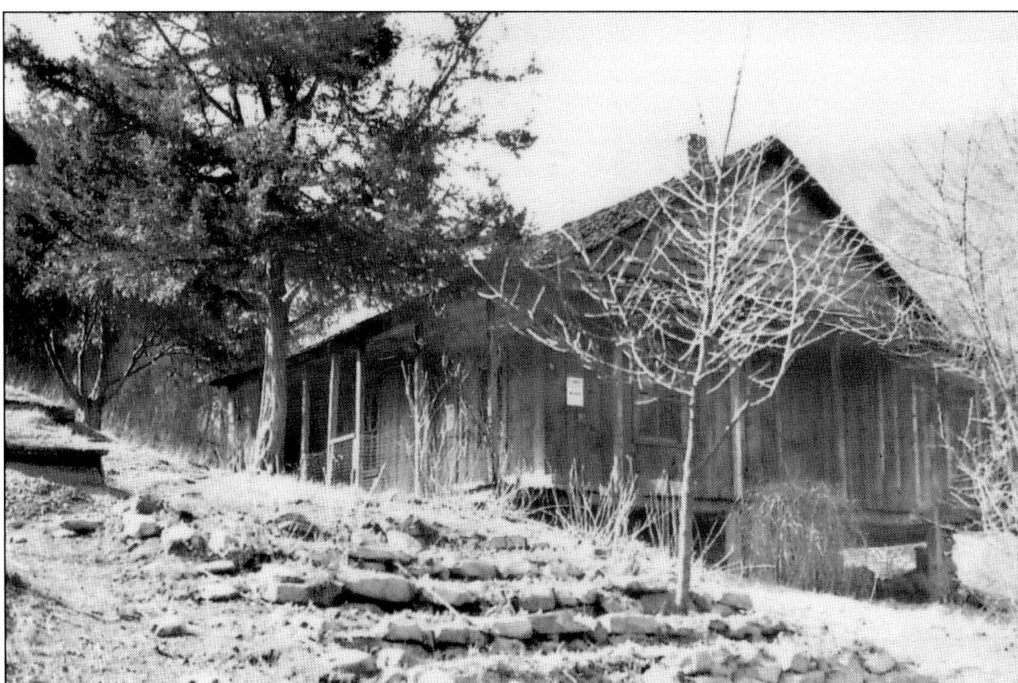

The home of George Wilson was located on Oliver Hollow Road in Carden's Bluff. (TVA photograph courtesy of Butler Museum.)

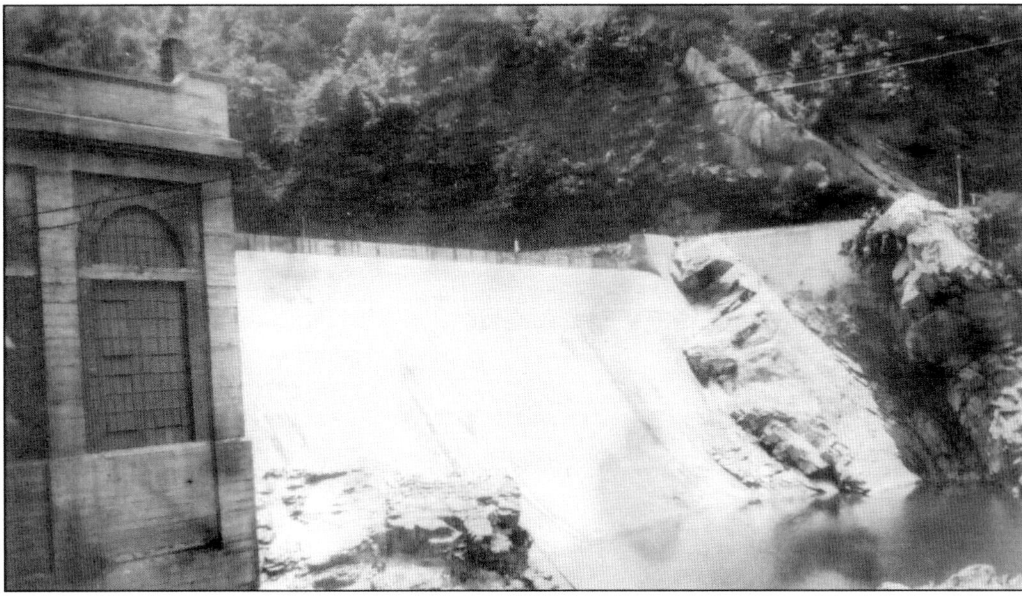

The Horseshoe Dam, another casualty of the 1940 flood, was the main source of power for the East Tennessee Light and Power Company. The dam was purchased, renovated, and renamed Wilbur Dam by TVA during the 1940s. (Photograph courtesy of George Walker.)

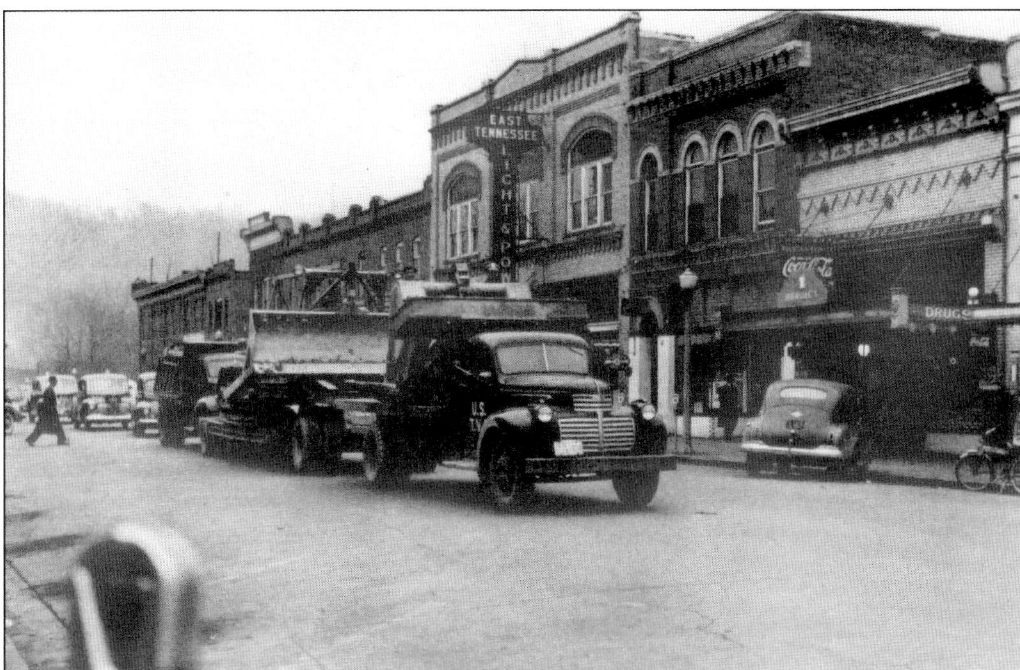

When the TVA came to a community, it purchased the local power boards in order to consolidate the cost of electrifying rural areas. Here a TVA work truck pauses in front of the newly acquired East Tennessee Light and Power Company in Elizabethton on its way to the Watauga Dam project site in 1942. (Photograph courtesy of Butler Museum.)

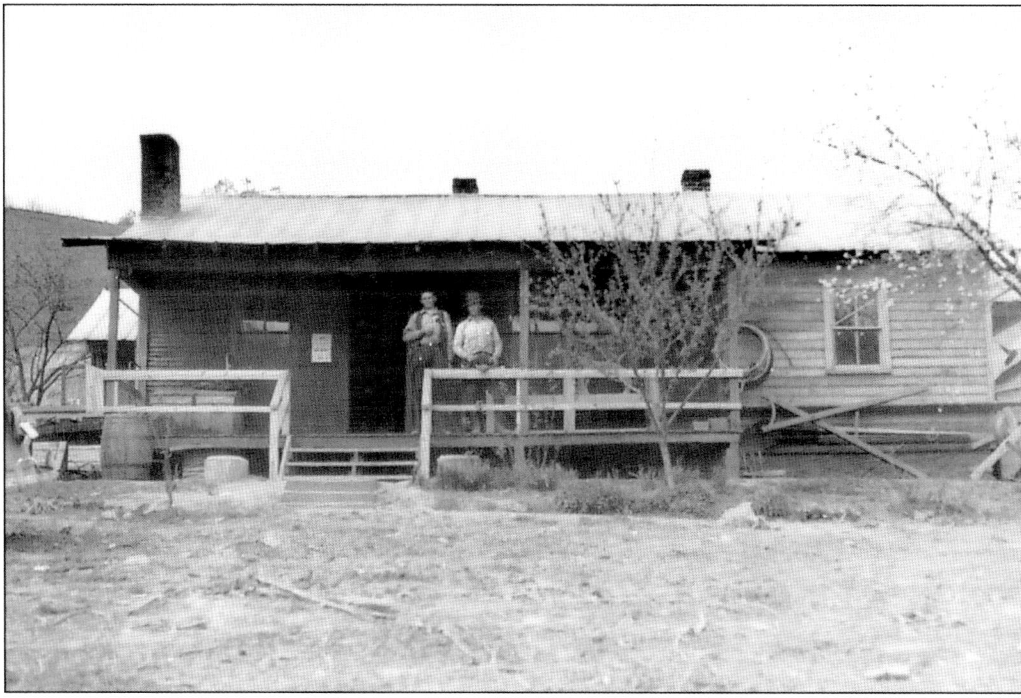

Pictured here is the headquarters of the Tennessee Valley Authority. Many young men of the Watauga Valley sought employment with the TVA. (TVA photograph courtesy of Butler Museum.)

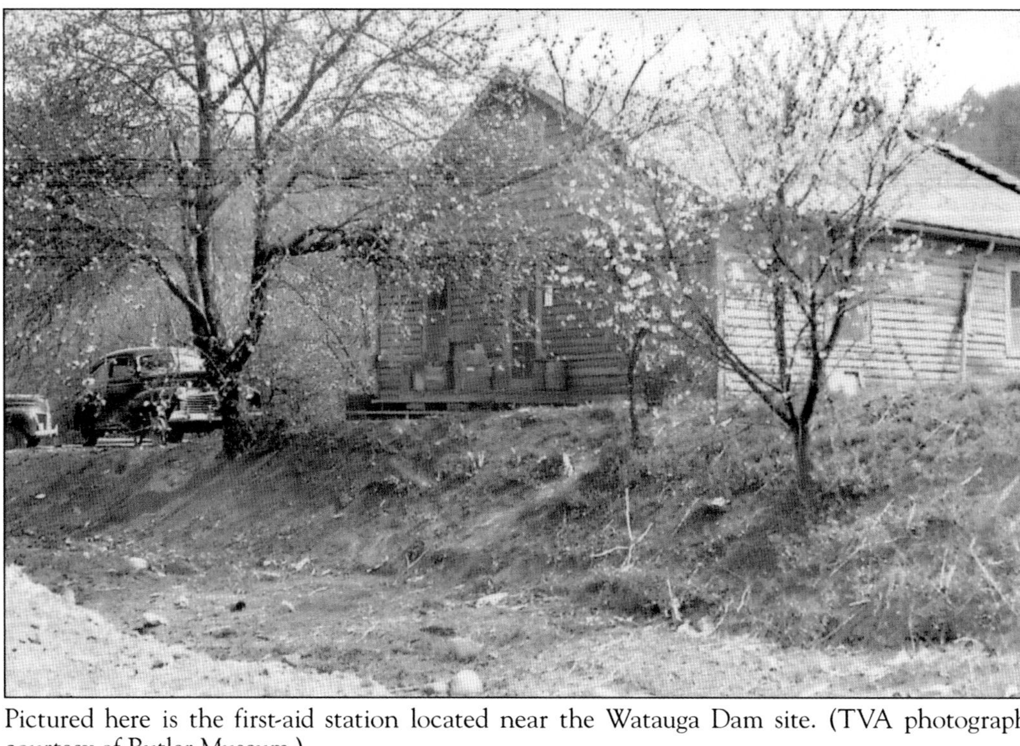

Pictured here is the first-aid station located near the Watauga Dam site. (TVA photograph courtesy of Butler Museum.)

Here TVA workers build the access road to the dam site before the Watauga River was diverted. (TVA photograph courtesy of Butler Museum.)

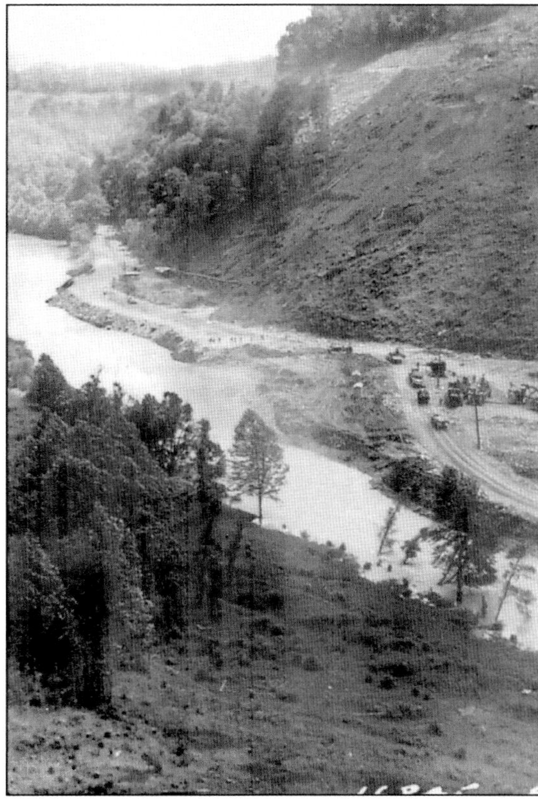

Workers blasted down to the bedrock of Carden's Bluff, as shown in this photograph, so that the dam would be as stable as possible. (TVA photograph courtesy of Butler Museum.)

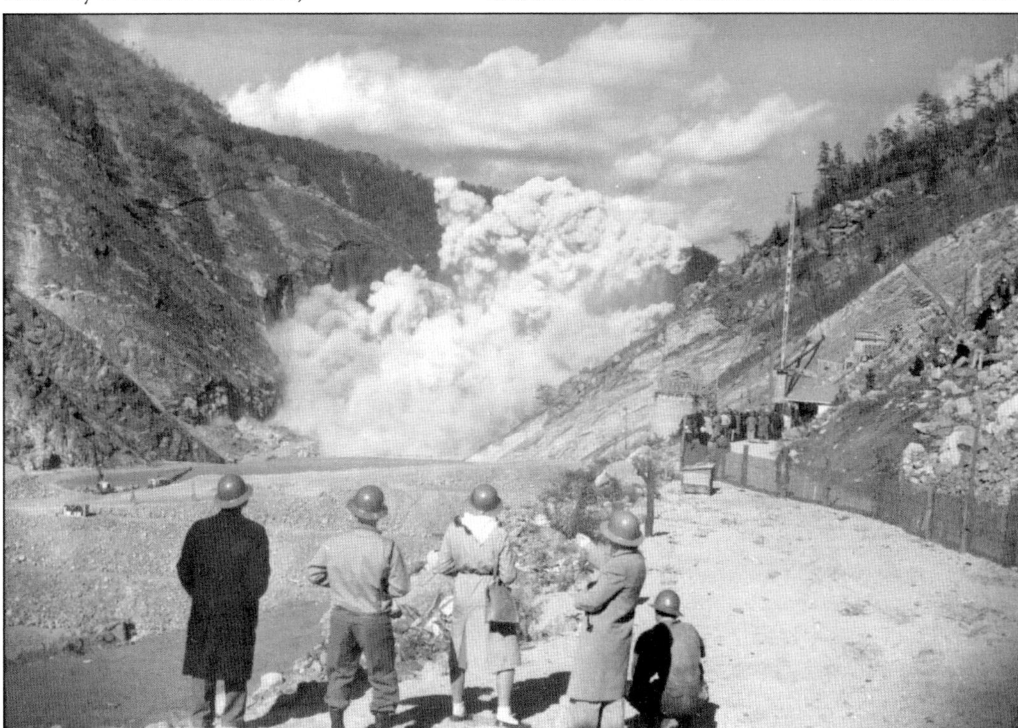

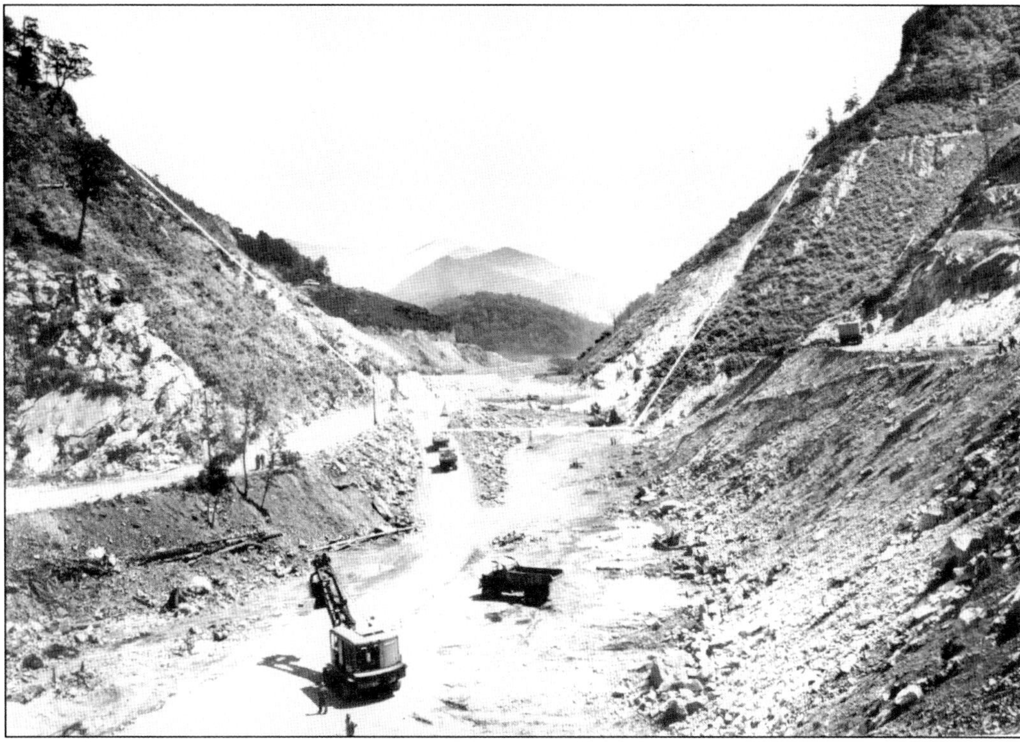

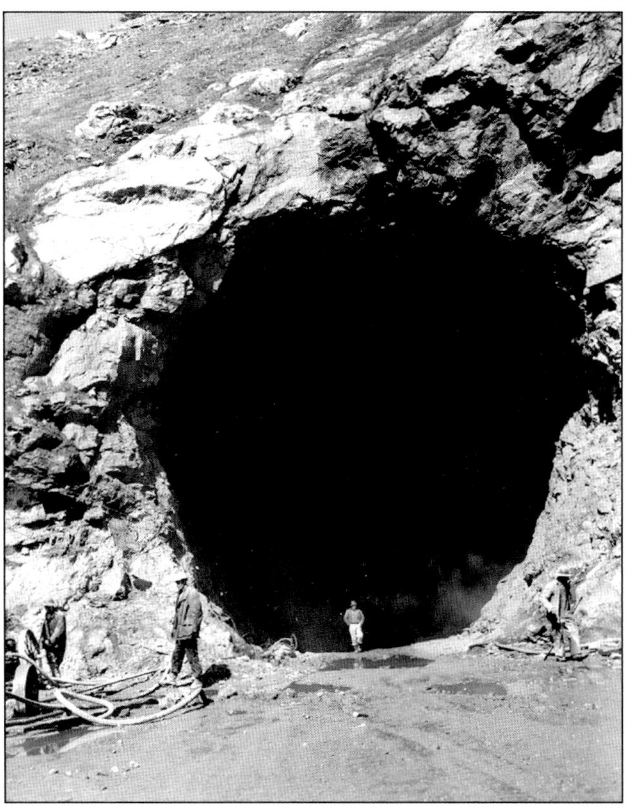

In this picture, work is progressing while the Watauga River is temporarily diverted. (TVA photograph courtesy of Butler Museum.)

Here a tunnel is cut into the mountain for the sluiceway, which allows water to return to the path of the river on the other side of the Watauga Dam. (TVA photograph courtesy of Butler Museum.)

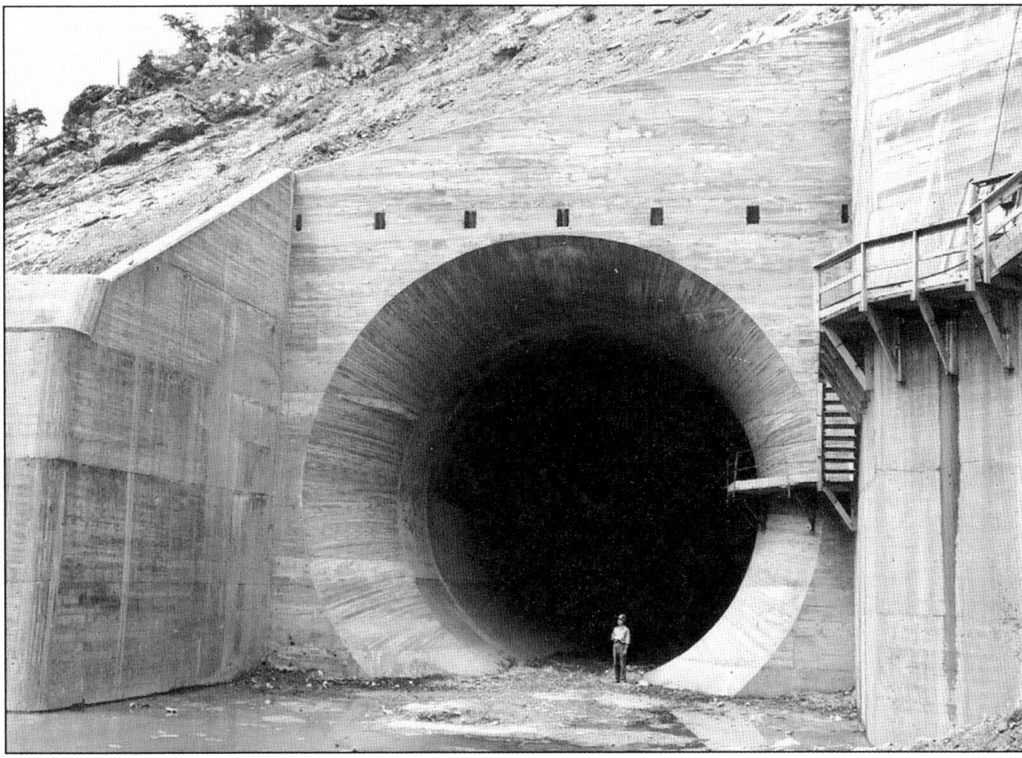

The sluiceway tunnel is shown here after concrete finishing. Most of the tunnel was finished before the War Production Board halted construction of the dam in 1942 so that resources could be used for the war effort. (TVA photograph courtesy of Butler Museum.)

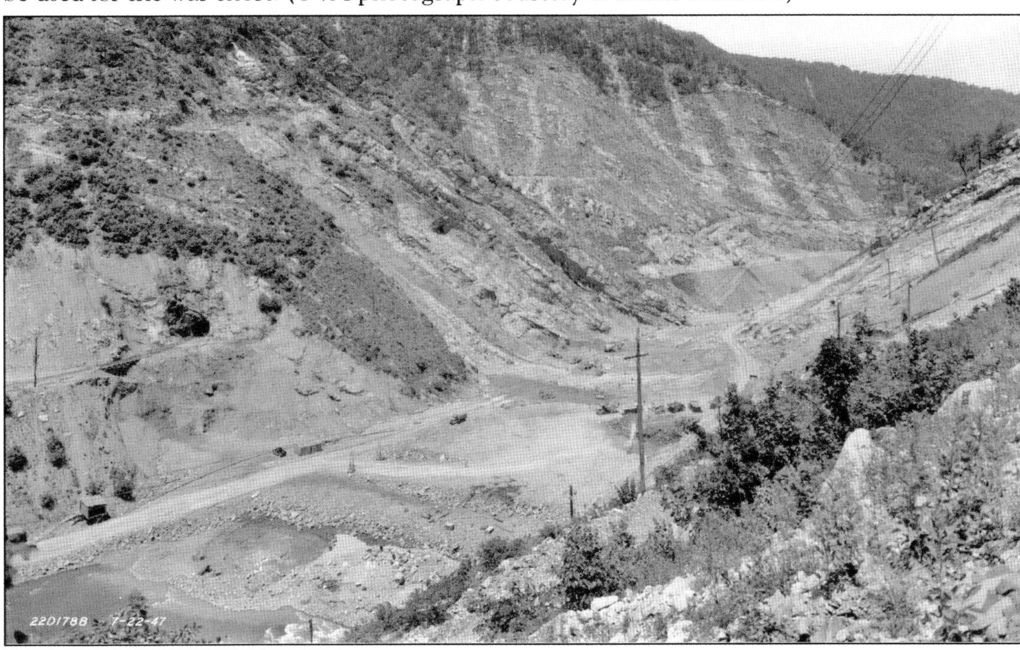

In 1946, work on the Watauga Dam resumed. This picture was taken in July 1947 as the first layers of the dam were laid. (TVA photograph courtesy of Butler Museum.)

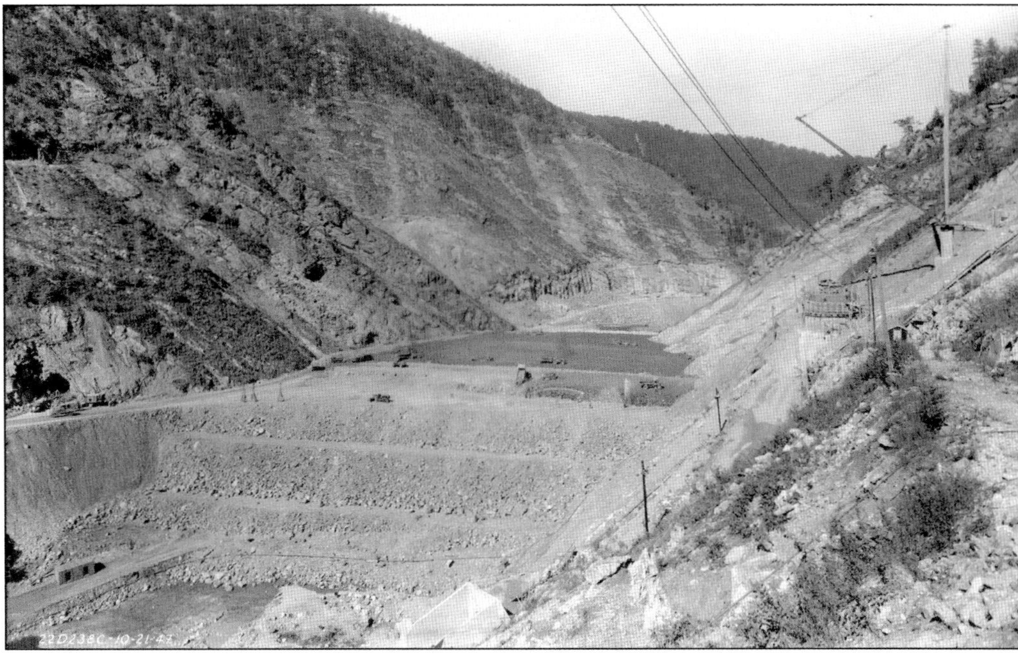

By October 1947, a second access road was built to reach the growing dam. Notice the men using the industrial water hoses to moisten the dirt to help pack down the earthen dam. At the bottom center, the sluiceway tunnel is shown with its concrete entranceway. Also notice the tunnel being cut on the right. This tunnel is the power-intake, which will funnel the water to the powerhouse. (TVA photograph courtesy of Butler Museum.)

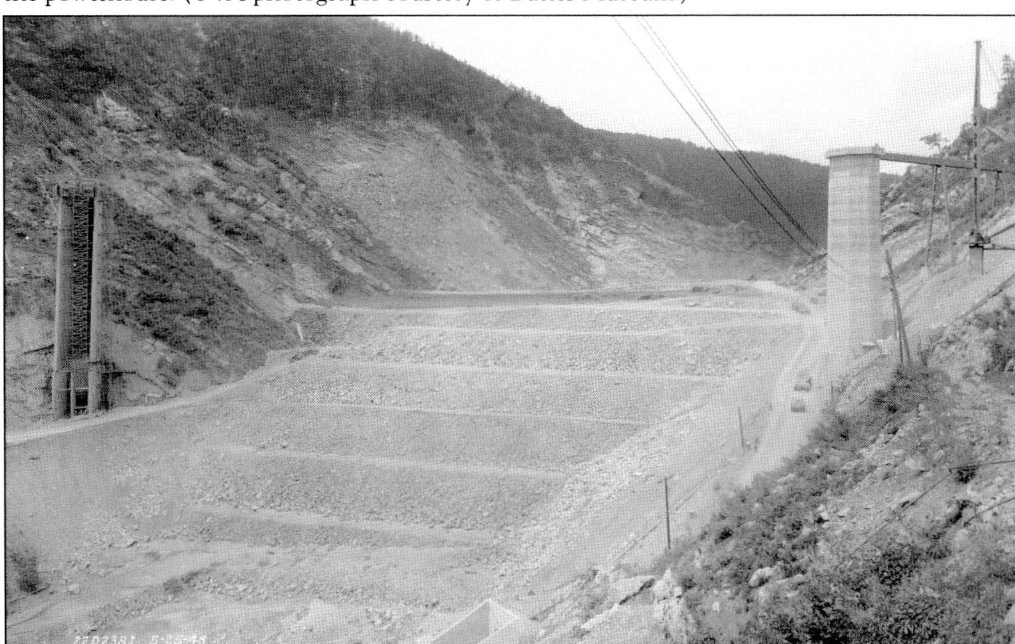

By May 1948, a third access road was cut to the new top of the dam as work continued. Notice the sluiceway tower on the right is complete and the power intake tower is also in place. (TVA photograph courtesy of Butler Museum.)

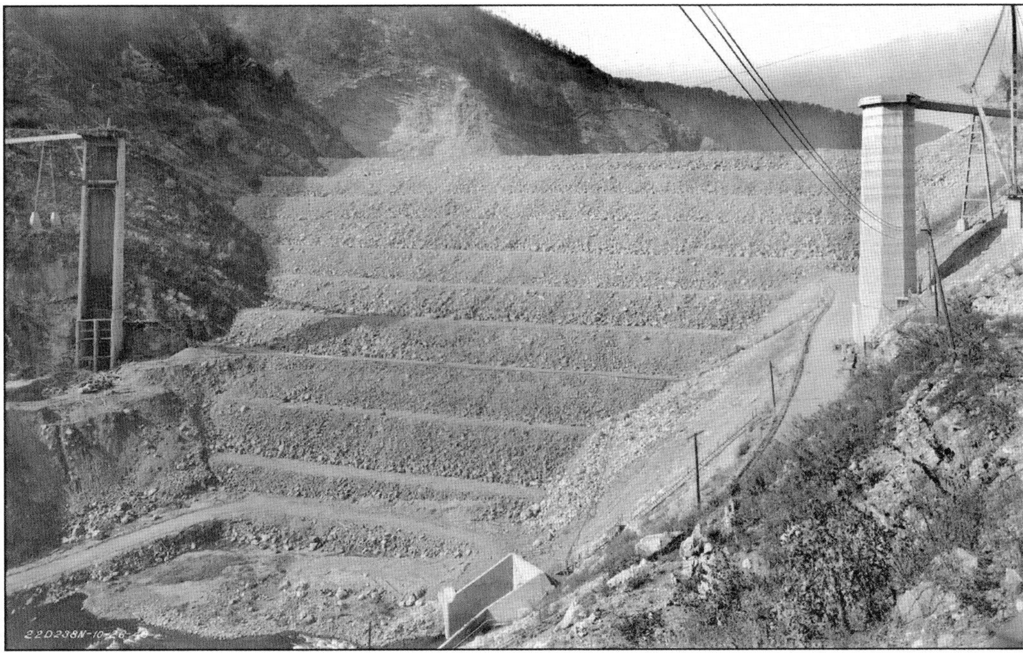

By November 1948, the Watauga Dam was finished and water was allowed to collect behind the dam. (TVA photograph courtesy of Butler Museum.)

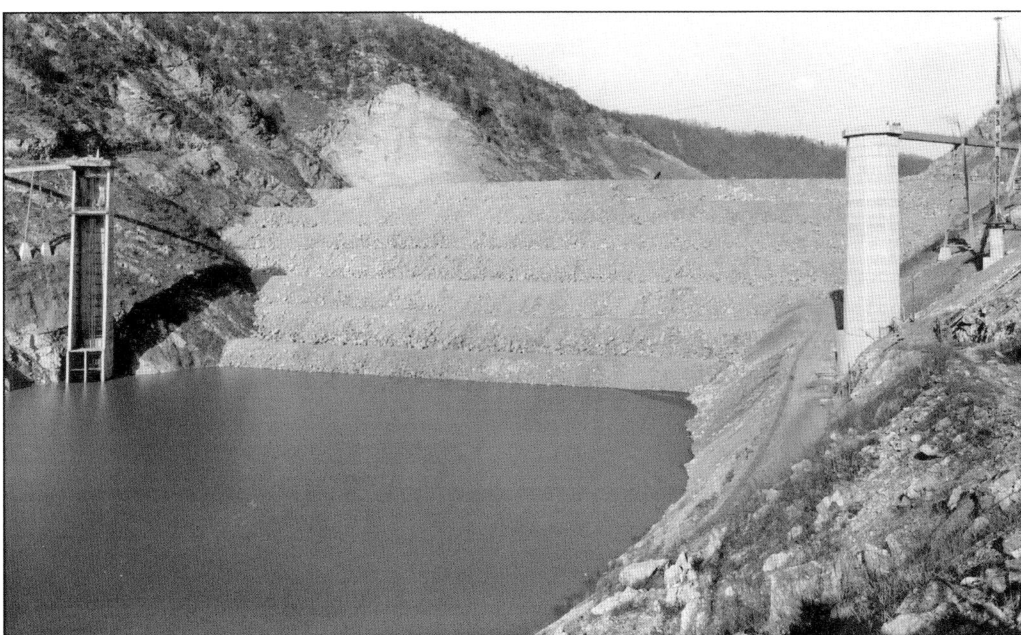

By December 1948, water had filled the reservoir much faster than anyone expected. The primary reason was because of the number of freshwater springs that fed the valley. (TVA photograph courtesy of Butler Museum.)

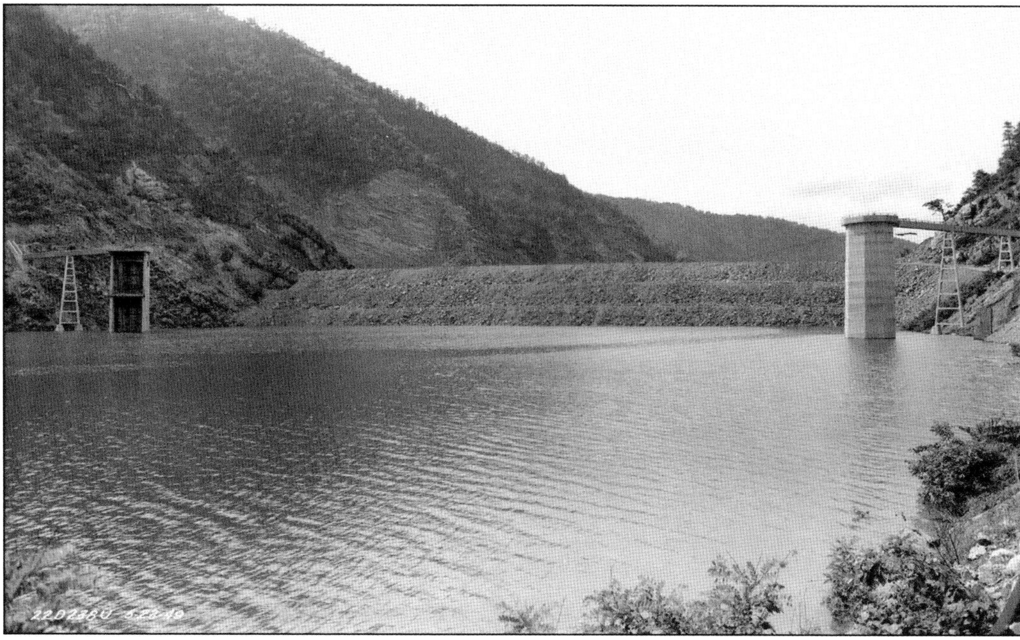

By May 1949, the Watauga Dam was fully functional and the Watauga Reservoir was full. (TVA photograph courtesy of Butler Museum.)

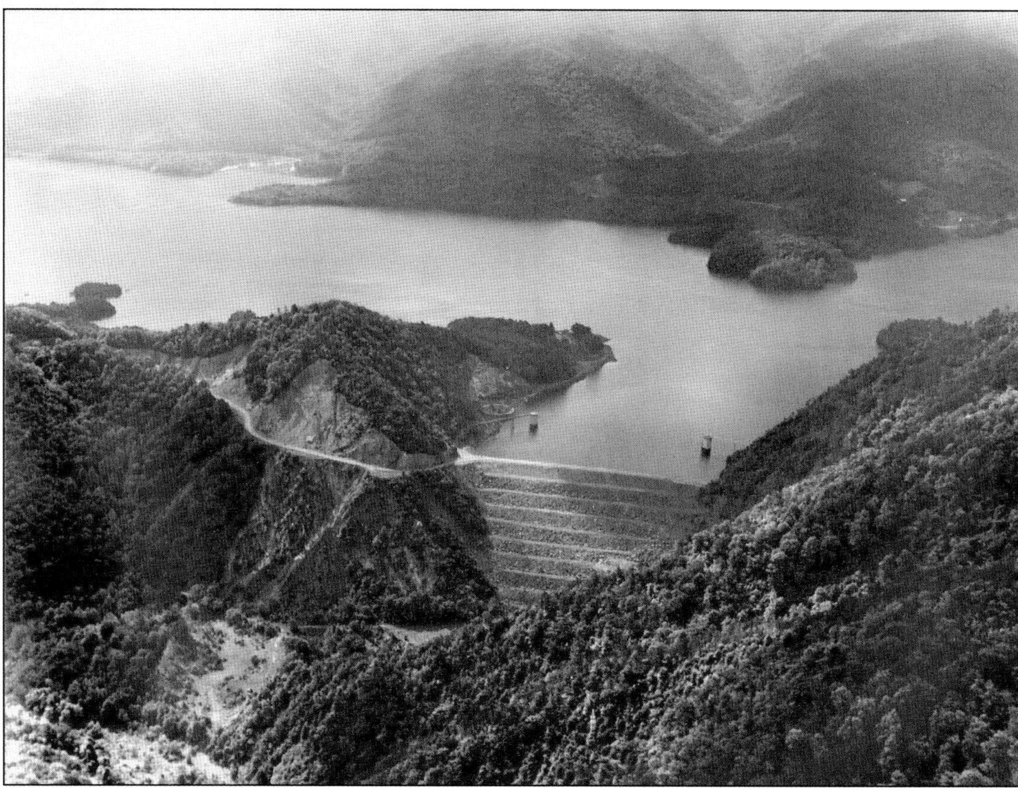

This aerial photograph emphasizes the immensity of the Watauga Dam and the enormous amount of water it holds back to create Watauga Lake. (TVA photograph courtesy of Butler Museum.)

Seven

Disappearing into the Mist

During the 1940s, residents of the Watauga Valley had to come to terms with the fact that the Tennessee Valley Authority would indeed build the Watauga Dam in Carden's Bluff—it was time to move. Individuals and families relocated to neighboring towns and communities such as Elizabethton, Hampton, Mountain City, or the newly developed community of Carderview (later renamed Butler). An auction was held as the old McQueen farm was divided up into lots and residents of Butler were encouraged to come and bid on sites for new homes. Some literally moved their homes, schools, and churches while others allowed their buildings to be demolished and chose to begin anew. More than 1,200 graves were exhumed and re-interred in cemeteries above the water line of the Watauga Reservoir. The community that had once thrived in the valley had now almost vanished into the mists of time.

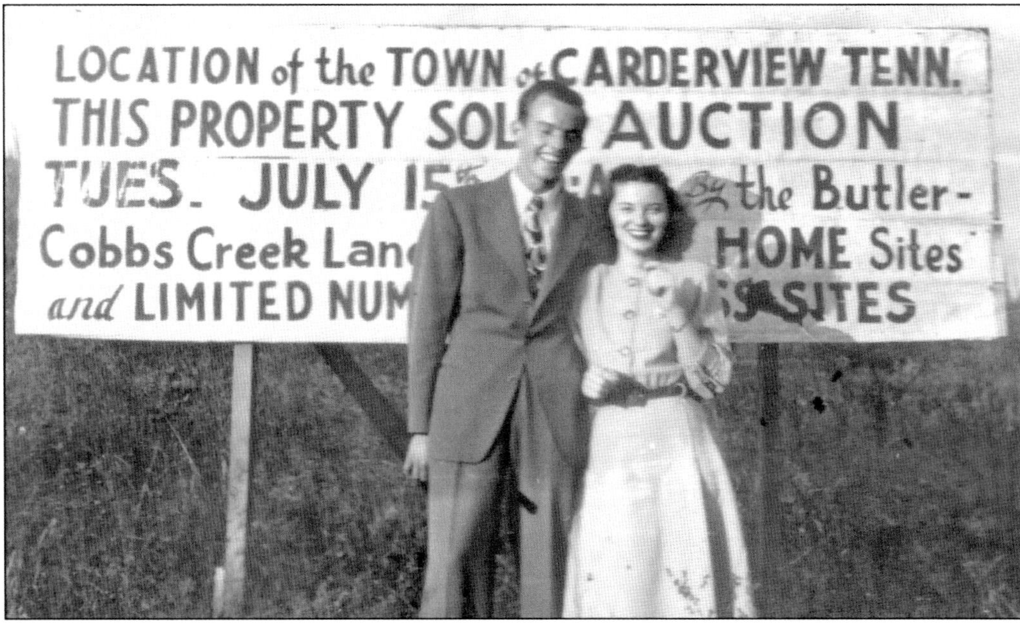

Elza Phillips (left) and Mary Walker stand in front of the sign advertising the auction held at McQueen's farm. (Photograph courtesy of Mary Walker Ward.)

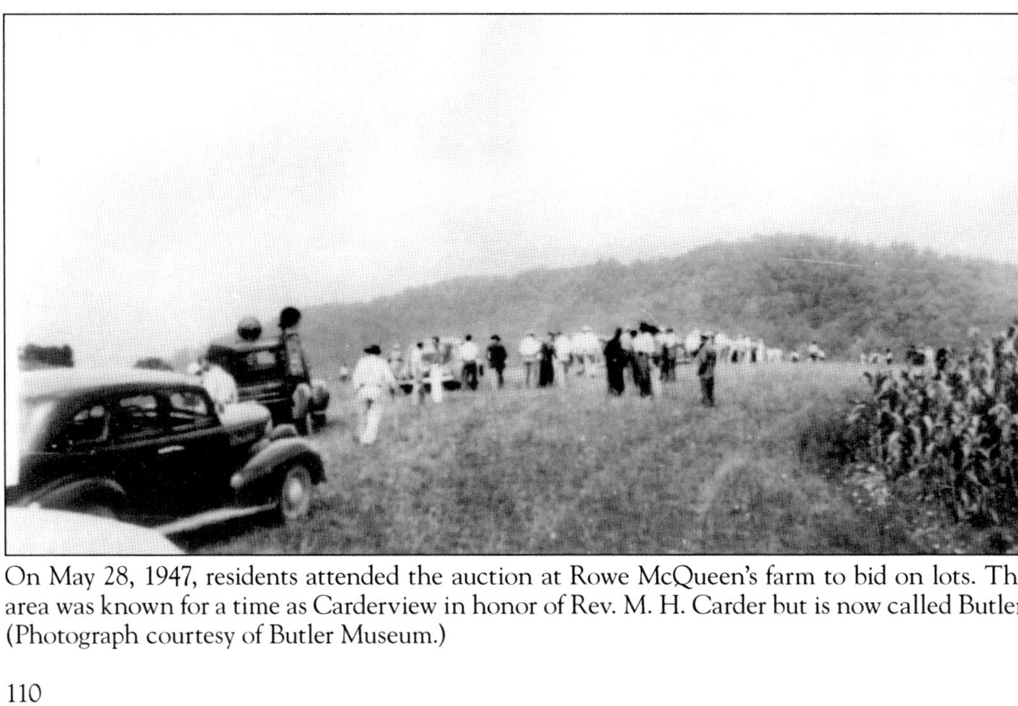

On May 28, 1947, residents attended the auction at Rowe McQueen's farm to bid on lots. The area was known for a time as Carderview in honor of Rev. M. H. Carder but is now called Butler. (Photograph courtesy of Butler Museum.)

On August 1, 1947, the Tennessee General Assembly passed a private act dissolving the city's charter and appointed Ira Q. Harris, pictured here, as the liquidating agent for Butler. (Photograph courtesy of James D. Robinson.)

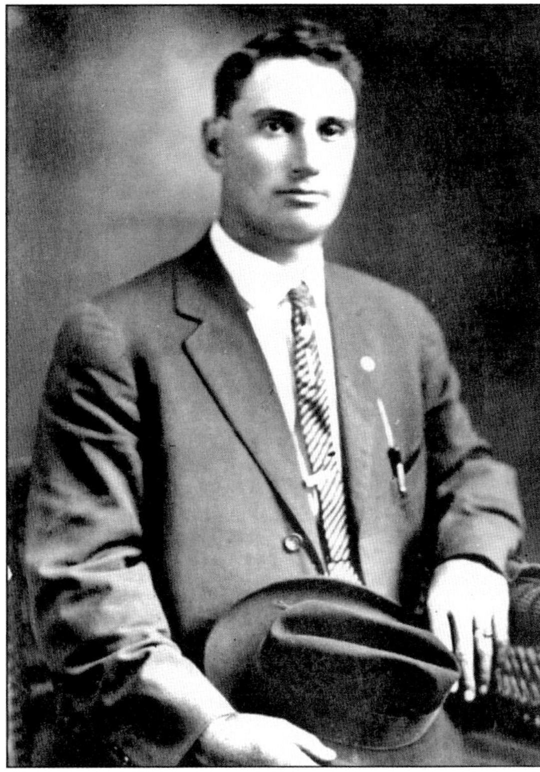

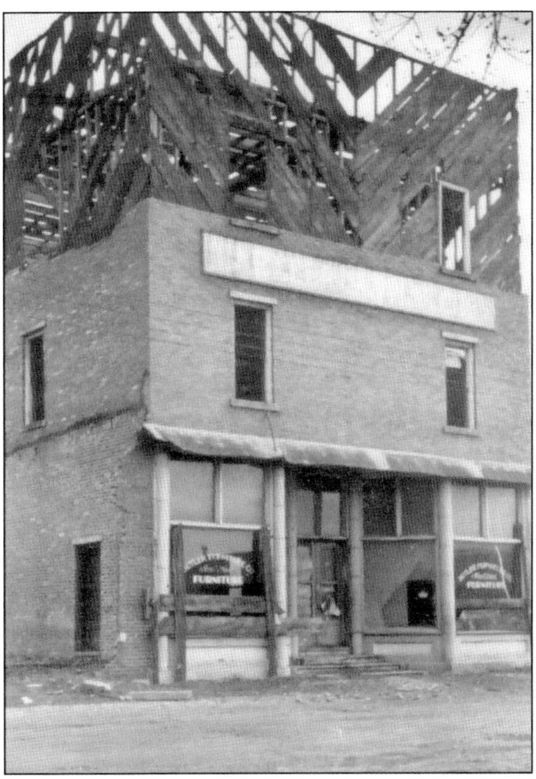

Ronda Goodwin purchased the building that once housed the Butler Furniture Company. He tore it down and used the brick to veneer his house, sidewalks, and steps in Doe Valley. (Photograph courtesy of Butler Museum.)

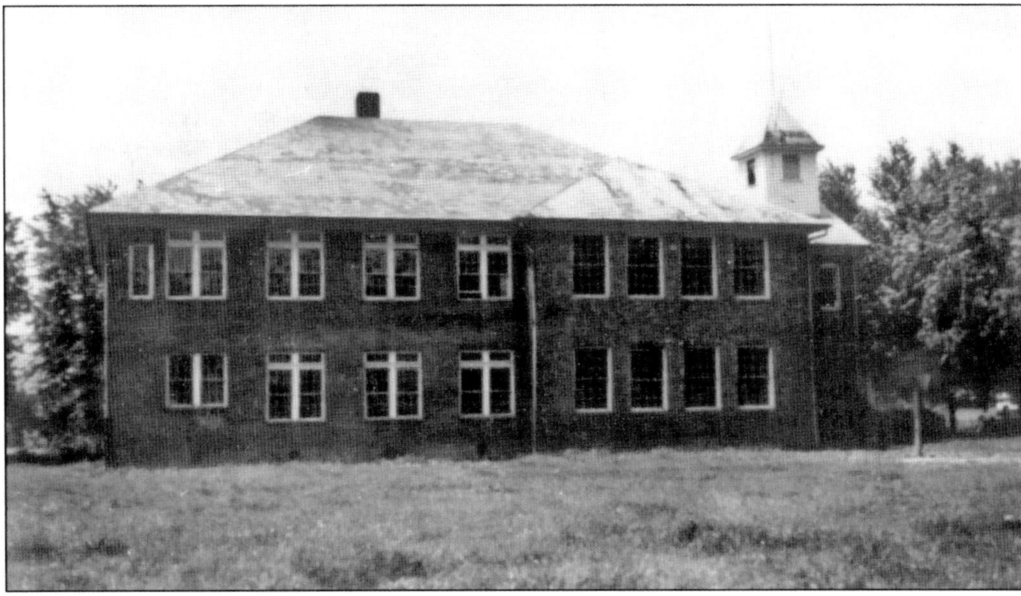

Butler City School was torn down. A new building was erected in Carderview in 1949. (Photograph courtesy of Butler Museum.)

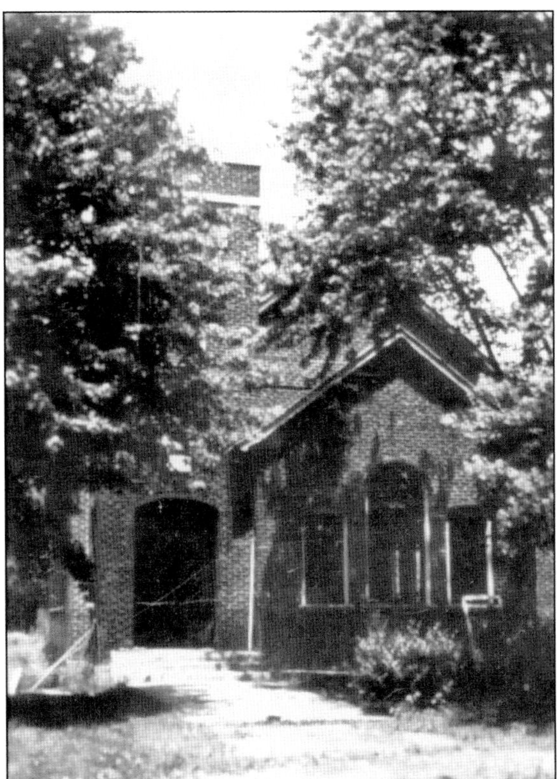

The Butler Christian Church was torn down by Rowe Phillips and his sons. The materials were used to build houses near Hampton. (TVA photograph courtesy of Butler Museum.)

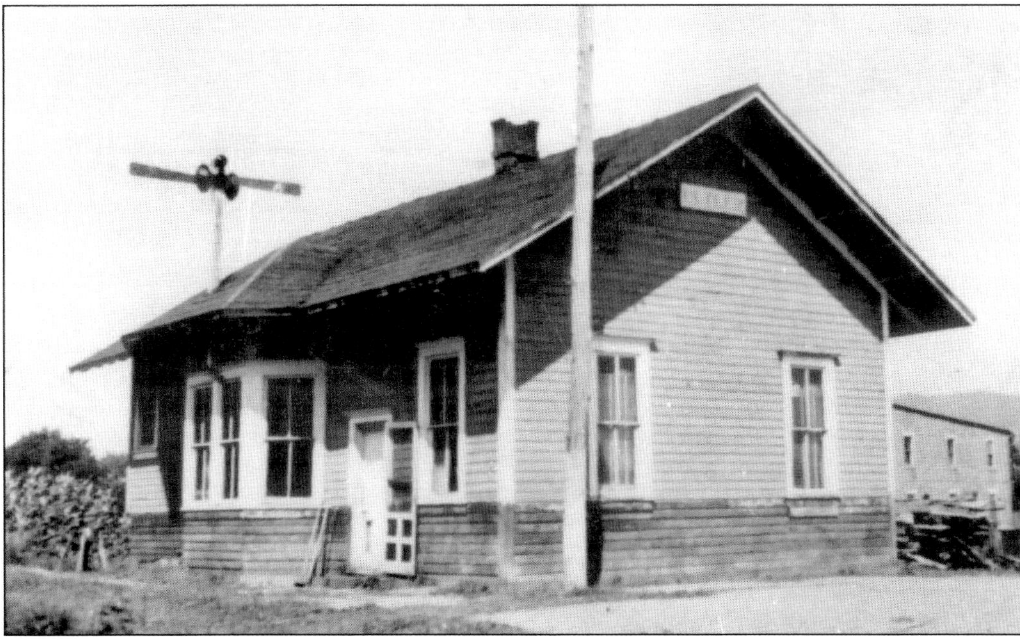

The Butler Depot was disassembled by the Barker family and reassembled in Virginia. (Photograph courtesy of Butler Museum.)

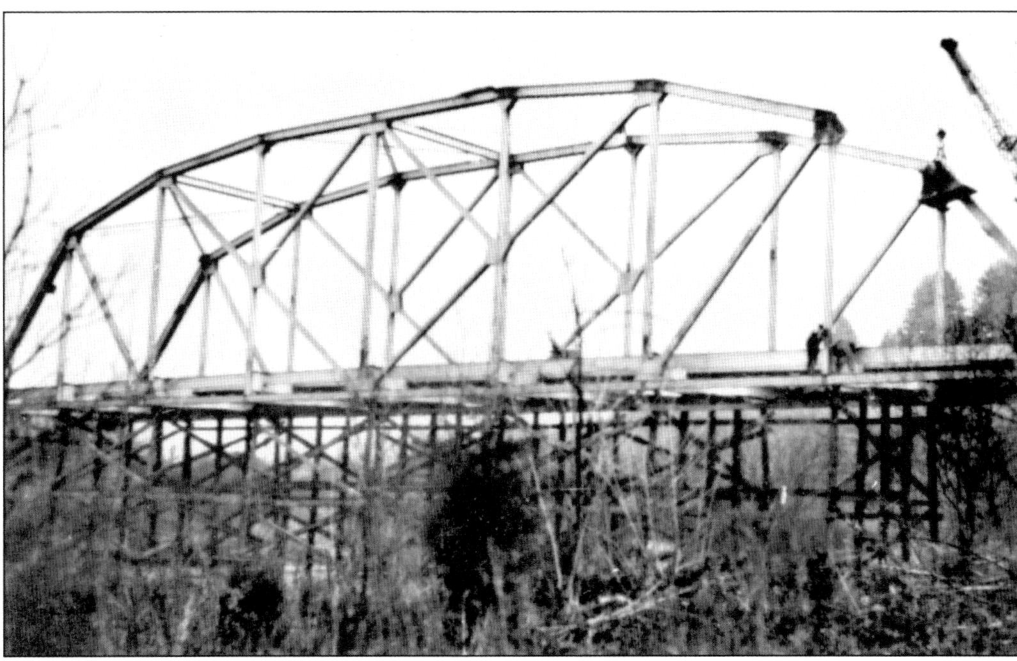

This photograph depicts the Butler Bridge as it is disassembled. (Photograph courtesy of Butler Museum.)

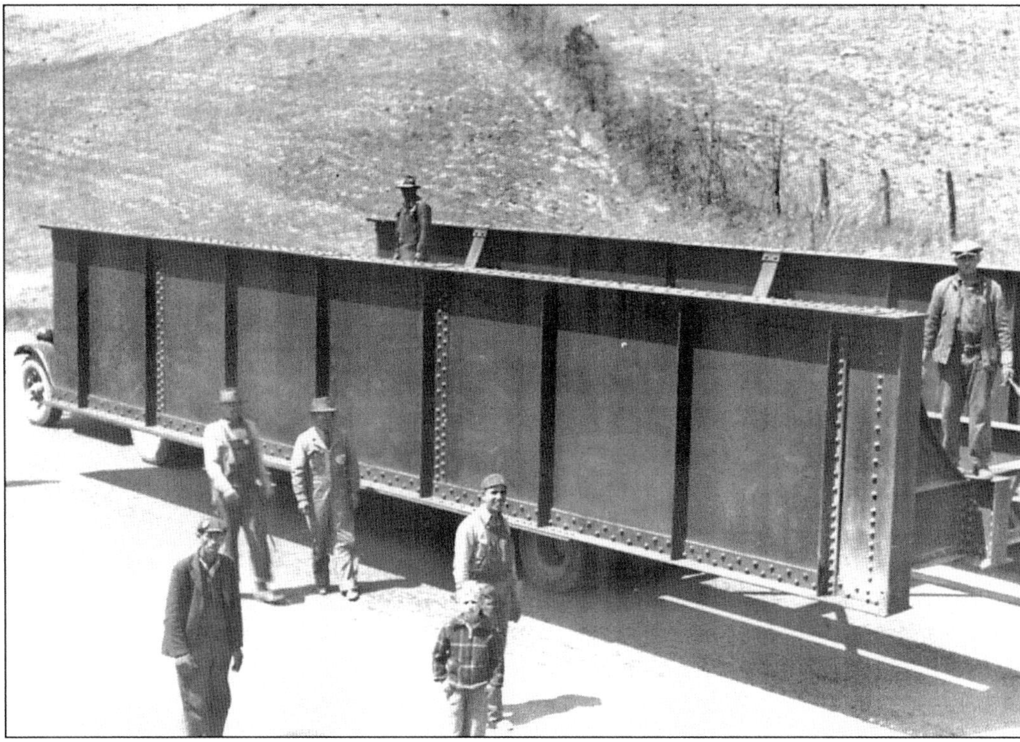

This railroad bridge is shown being removed from Butler for use at Holston. (Photograph courtesy of Butler Museum.)

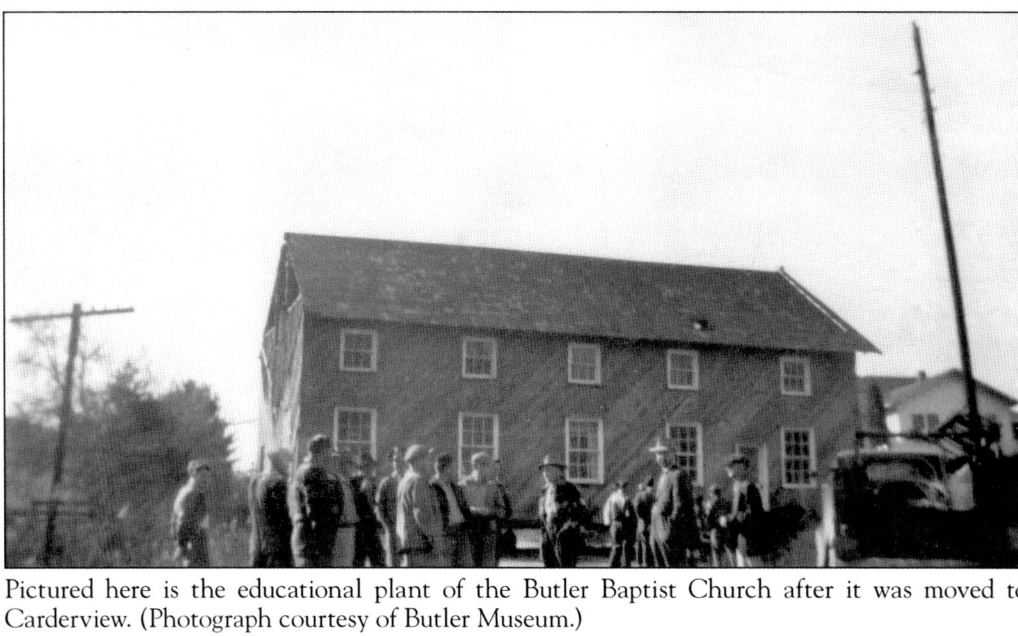

Pictured here is the educational plant of the Butler Baptist Church after it was moved to Carderview. (Photograph courtesy of Butler Museum.)

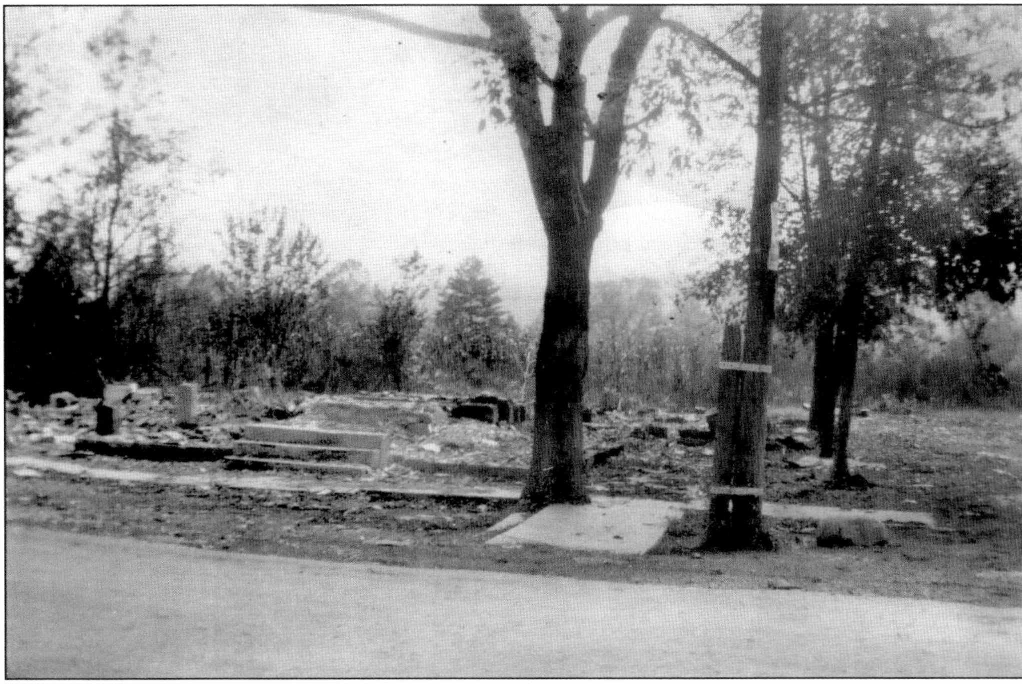

The foundation and steps are all that remain of the Blue Bird Tea Room and Bus Station. (Photograph courtesy of Butler Museum.)

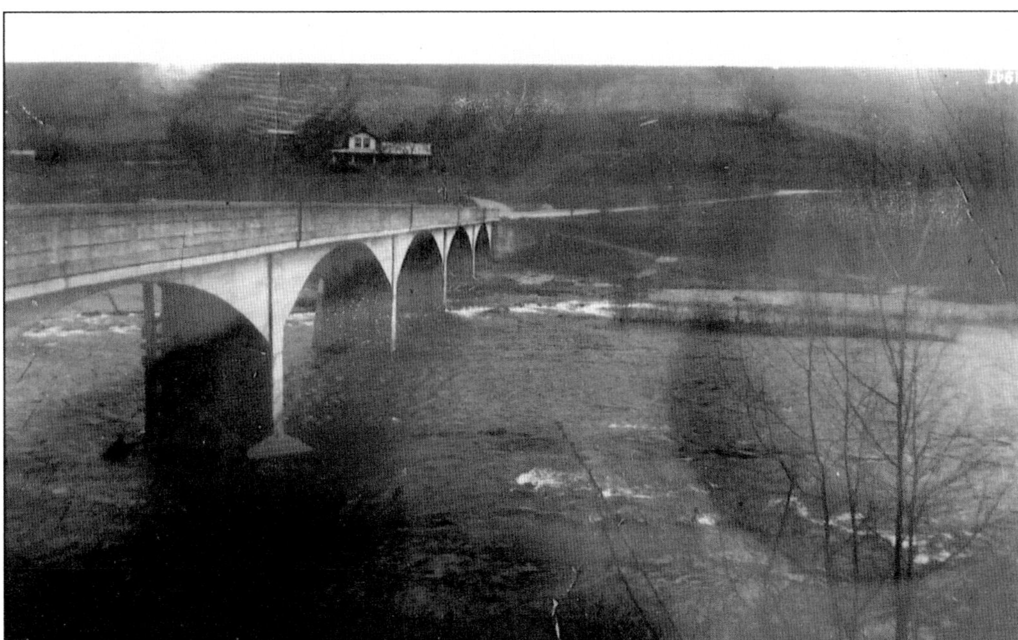

The Fish Springs Bridge, which at one time provided access across the Watauga River, was not torn down but remains intact beneath Watauga Lake. (Photograph courtesy of Butler Museum.)

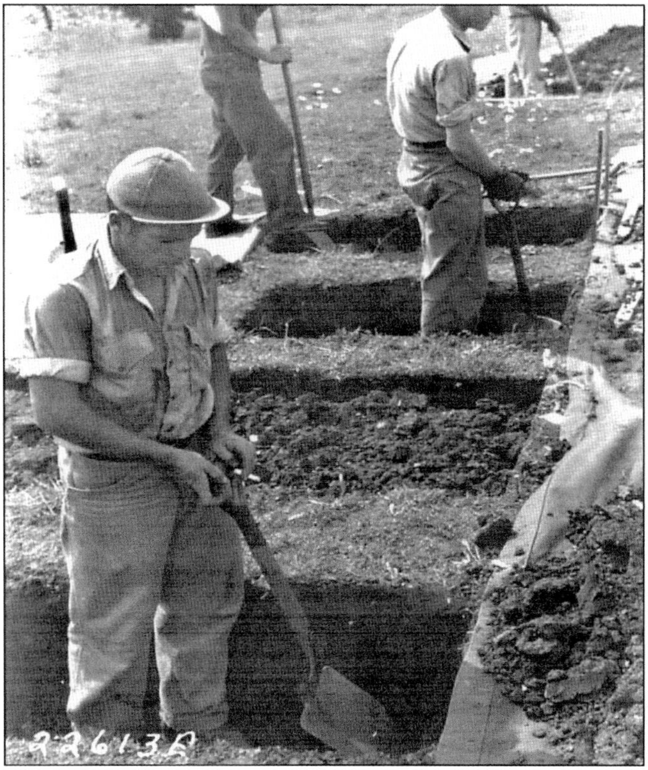

Workers are depicted here working on the re-interment project that moved corpses from Butler to Caldwell Springs. (Photograph courtesy of Butler Museum.)

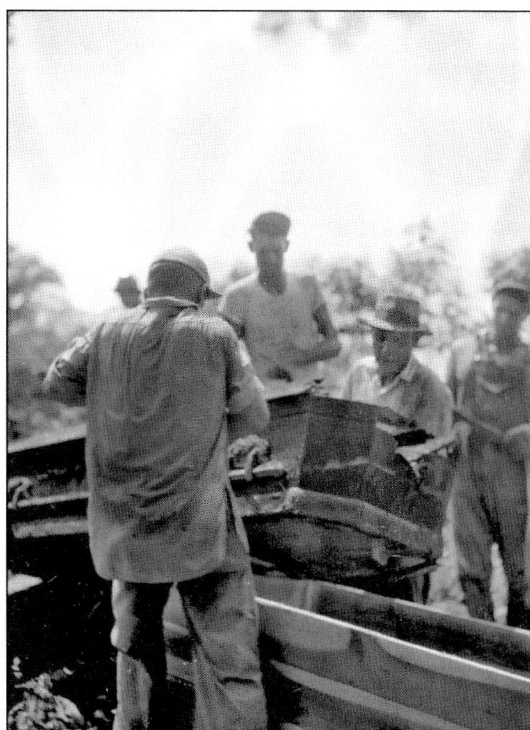

Workers load a casket into a wooden box that will be transported to a cemetery outside of the Watauga Reservoir. (Photograph courtesy of Butler Museum.)

The home of Fred M. Matherly is shown as it is moved on Cobbs Creek Road in July 1948. Notice Highway 67, which provided access into the new community of Carderview, being built in the background. (Photograph courtesy of Butler Museum.)

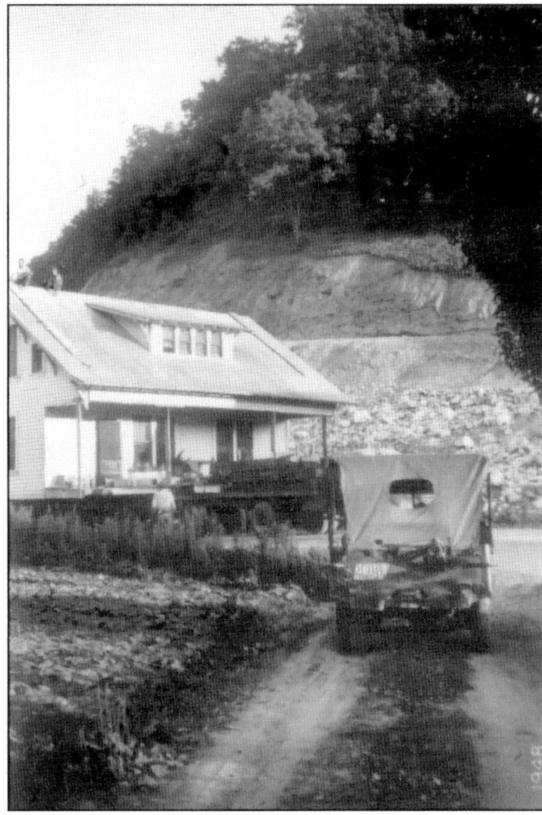

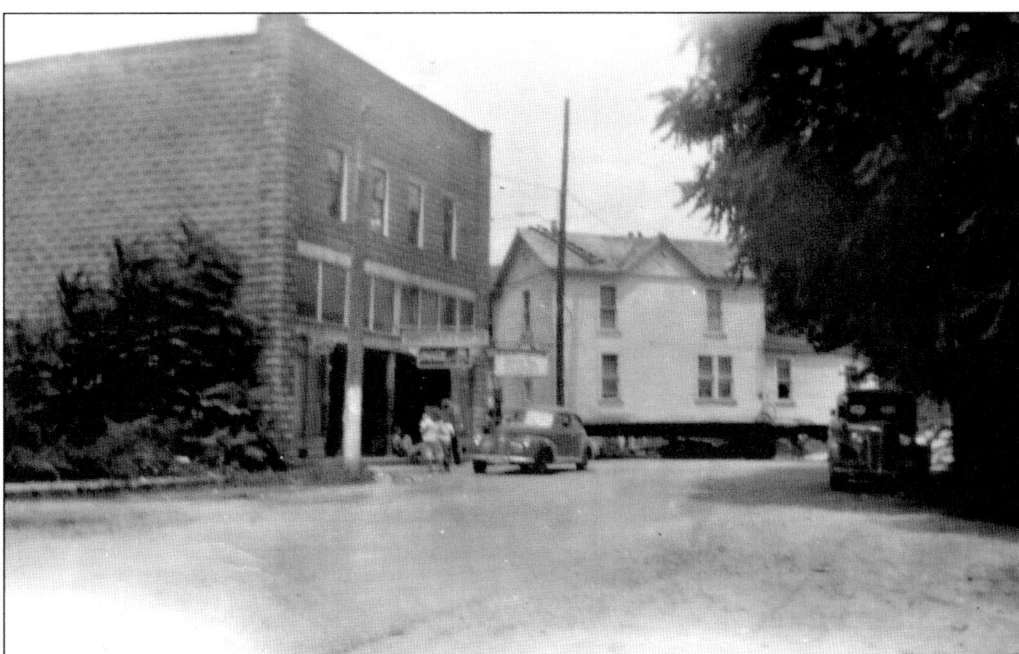

In this photograph, R. T. Smith's house, which housed the telephone office during the 1920s, is being driven from Main Street onto Spring Street. (Photograph courtesy of Butler Museum.)

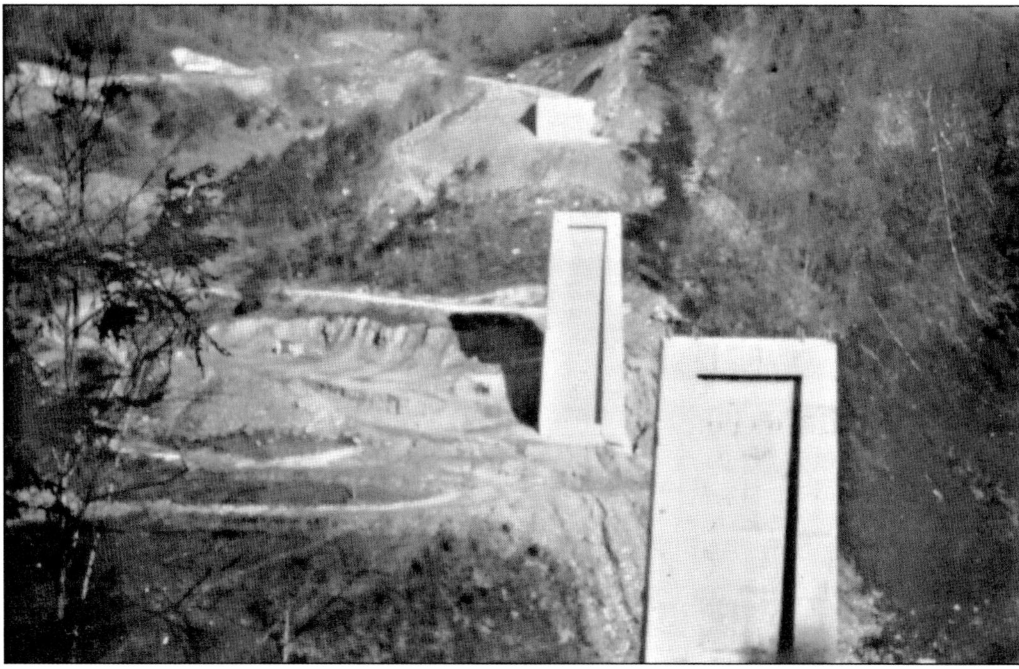

Butler Memorial Bridge, originally called the Dale Neely Bridge, was under construction in 1947. (Photograph courtesy of Butler Museum.)

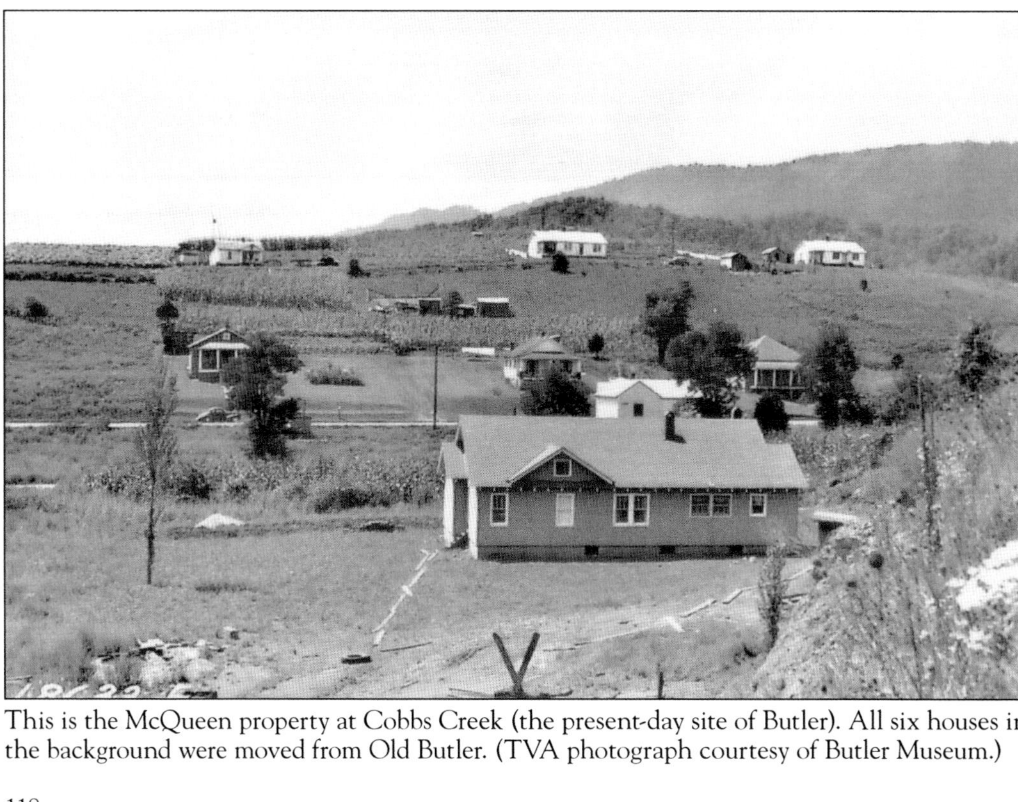

This is the McQueen property at Cobbs Creek (the present-day site of Butler). All six houses in the background were moved from Old Butler. (TVA photograph courtesy of Butler Museum.)

Eight

Butler Revisited

Even though the site of the city lies beneath the Watauga Lake, the memory of Old Butler lives on. Each year in August, the Watauga Academy Alumni Association holds a reunion. In 1985, Bill Trivette wrote and recorded a song called "Our Little Town: Butler Revisited," which commemorated the town and its former residents. In 1999, the East Tennessee State University Division of Theatre and the Butler Museum Committee presented the play *River Rising: TVA and the Town of Butler.* The play, written and adapted by Bobby Funk and directed by Ron McIntyre-Fender, was a dramatization based on a series of interviews of former Old Butler residents. To maintain the earthen dam, the Tennessee Valley Authority periodically (every 30 to 35 years) must drain the Watauga Reservoir. In 1983, former residents and other interested individuals from the region were given the opportunity to revisit the streets of Old Butler. In 2000, the Butler Museum, a facility located in New Butler, opened to provide the opportunity for the region's people to "revisit" Old Butler any time they desired.

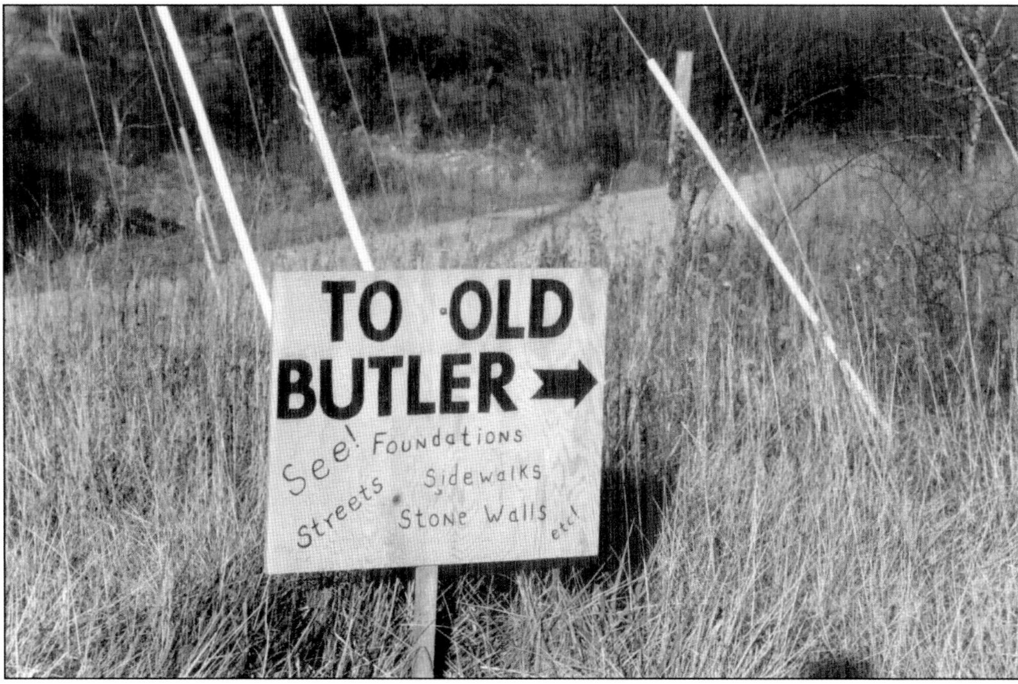

In December 1983, when the Watauga Reservoir was emptied for dam maintenance, this sign pointing to Old Butler beckoned visitors to once again walk along the streets and visit old sites that had been covered by Watauga Lake for 35 years. (Photograph courtesy of Butler Museum.)

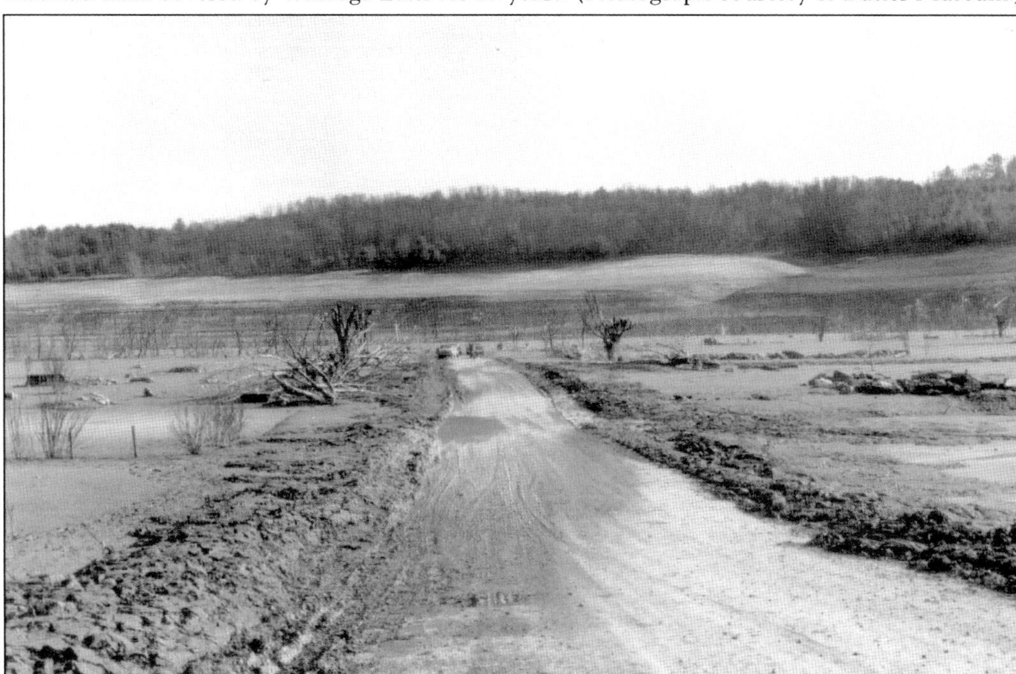

This photograph, taken of Spring Street toward Main Street, shows the streets that were cleared of mud and silt by the Johnson County Highway Department for visitors. (Photograph courtesy of Butler Museum.)

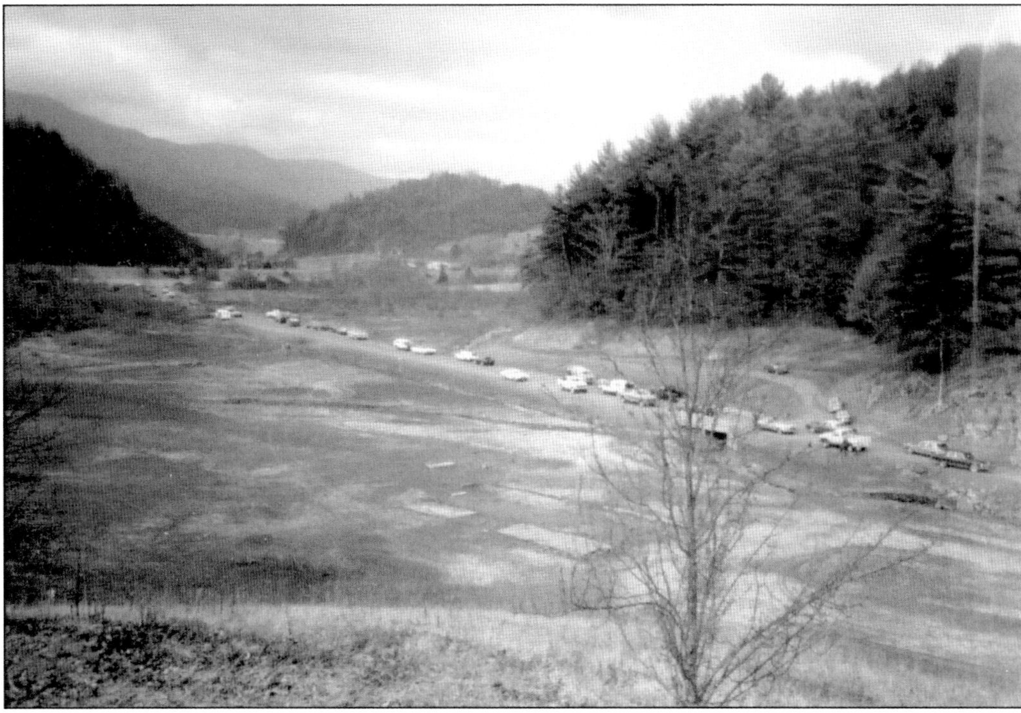

Pictured here are cars driving along Cobbs Creek Road and going down into Old Butler. (Photograph courtesy of Butler Museum.)

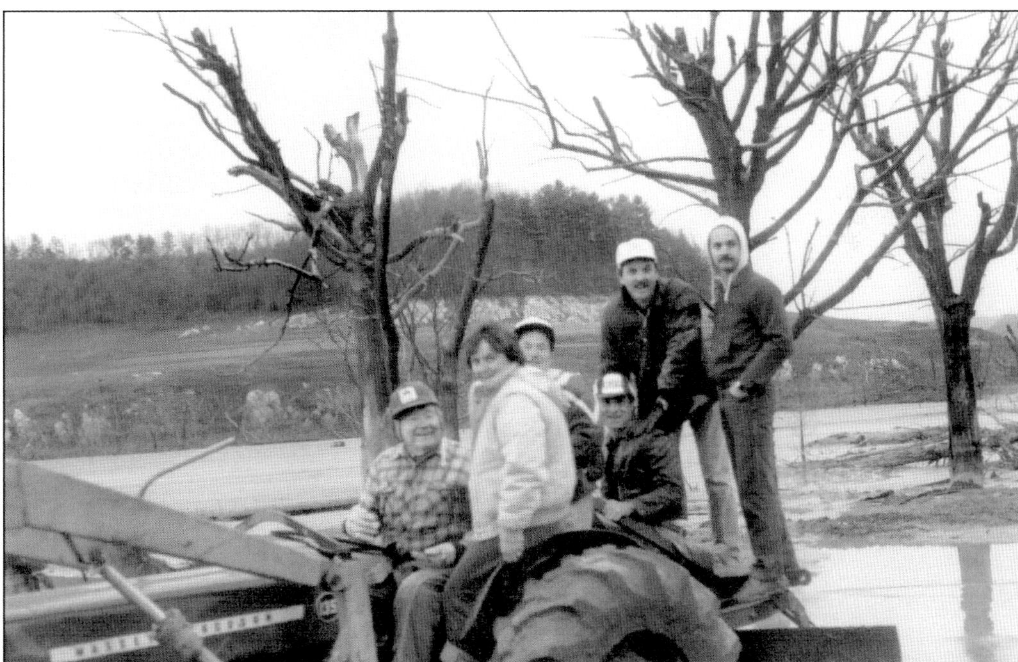

Here Joe McQueen transports visitors to Old Butler on his tractor. (Photograph courtesy of Butler Museum.)

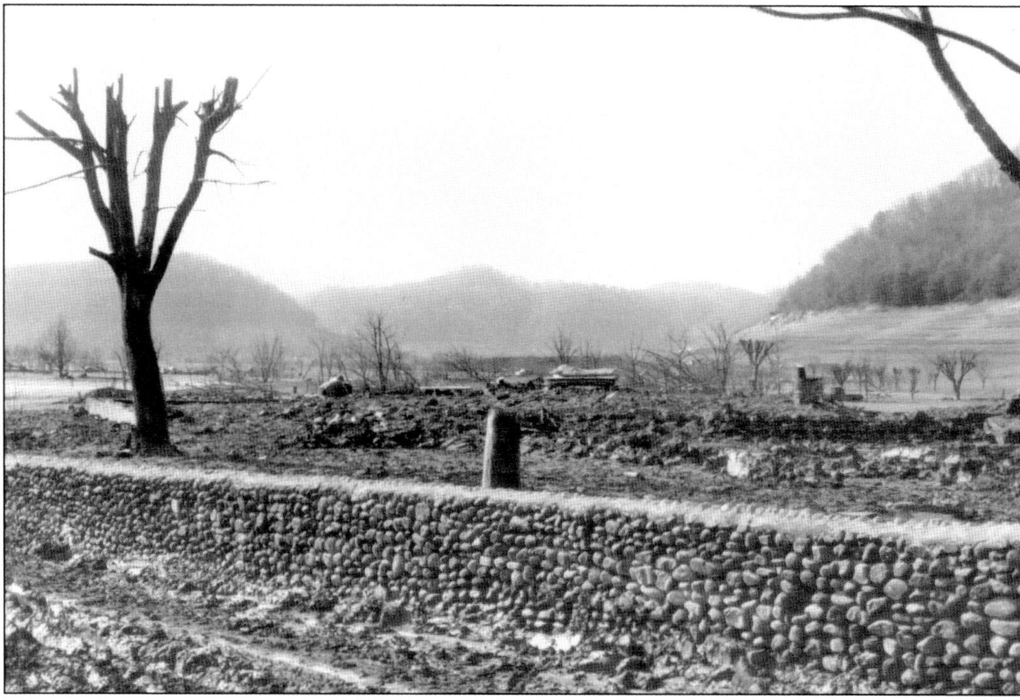

This photograph was taken near the site where Butler Baptist Church once stood. Notice the stone wall that still borders the property. (Photograph courtesy of Butler Museum.)

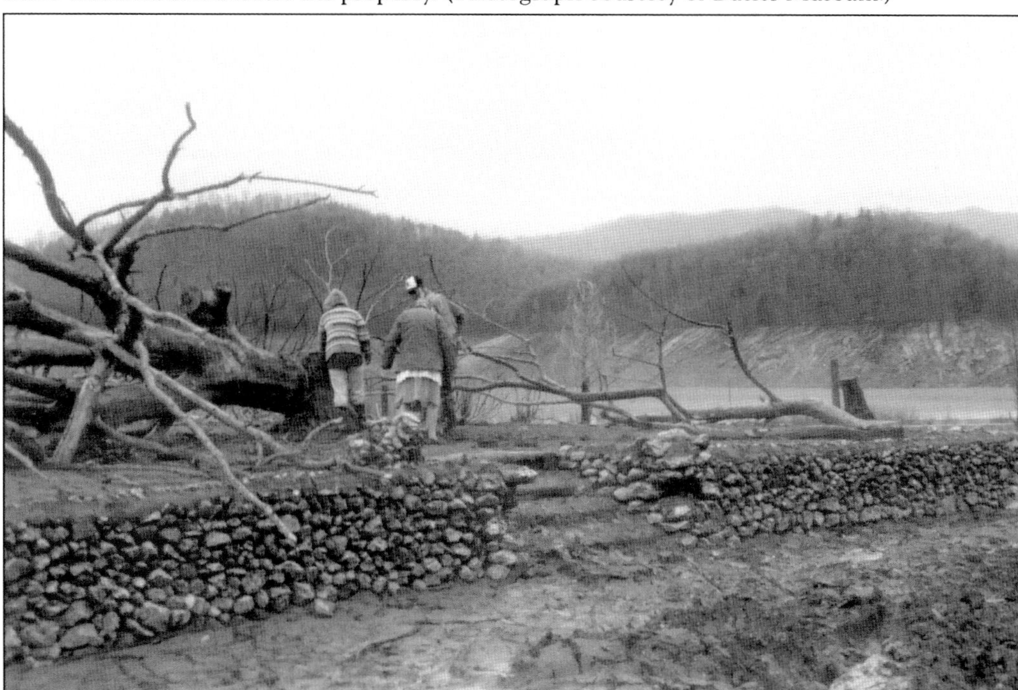

During the lake draining, archeologists dug for artifacts. Some artifacts were thousands of years old; however, these searchers at the site of the Frank Goodwin home were looking for artifacts that were only decades old. (Photograph courtesy of Butler Museum.)

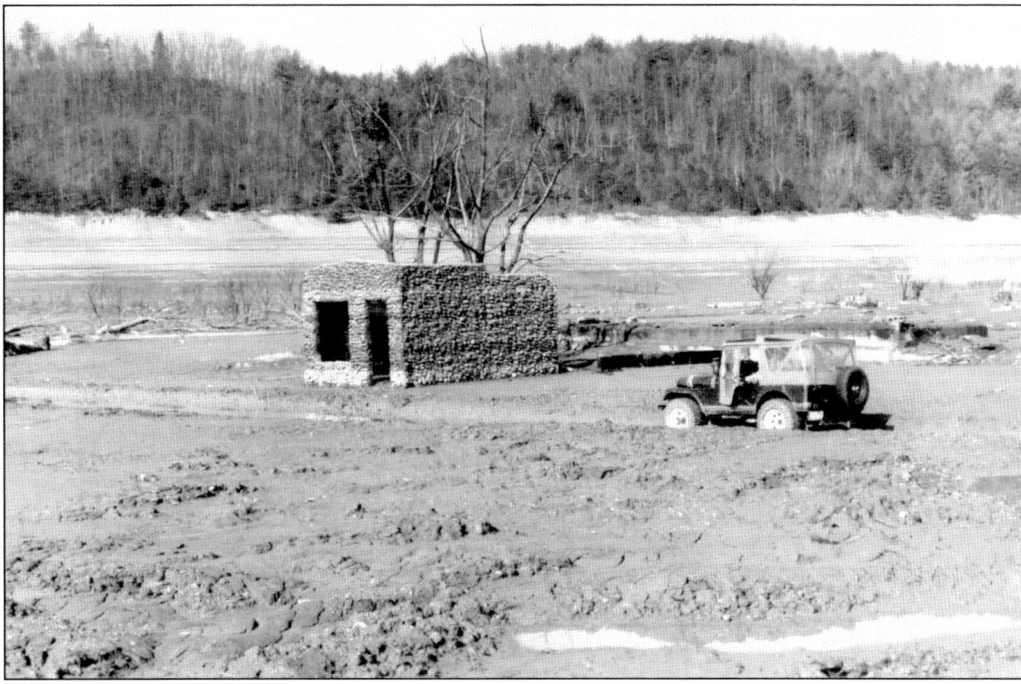

The building of the City Shoe Shop, one of the only buildings that was not torn down or relocated when the Watauga Dam was built in the 1940s, still stands beneath the Watauga Lake as a silent reminder of days gone by. (Photograph courtesy of Butler Museum.)

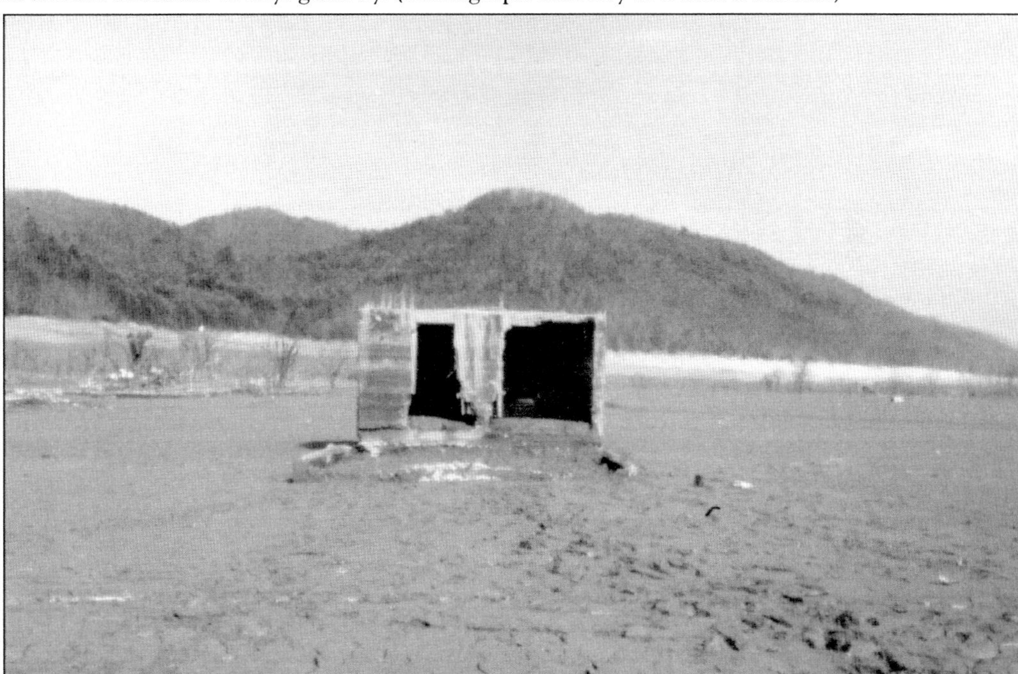

The empty cells of the jail, once locally known as the "Bug House," still stand after being covered by the waters of the Watauga Lake for 35 years. (Photograph courtesy of Butler Museum.)

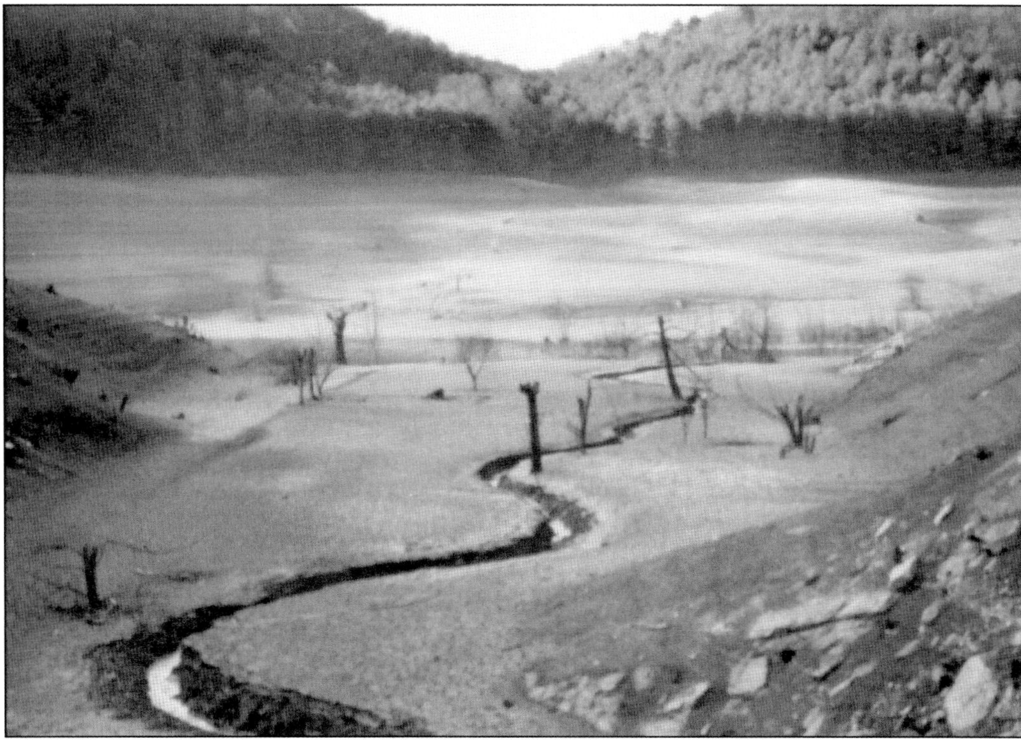

This photograph of Cobbs Creek depicts how nature reclaims her bounds, even after being diverted for 35 years. (Photograph courtesy of Butler Museum.)

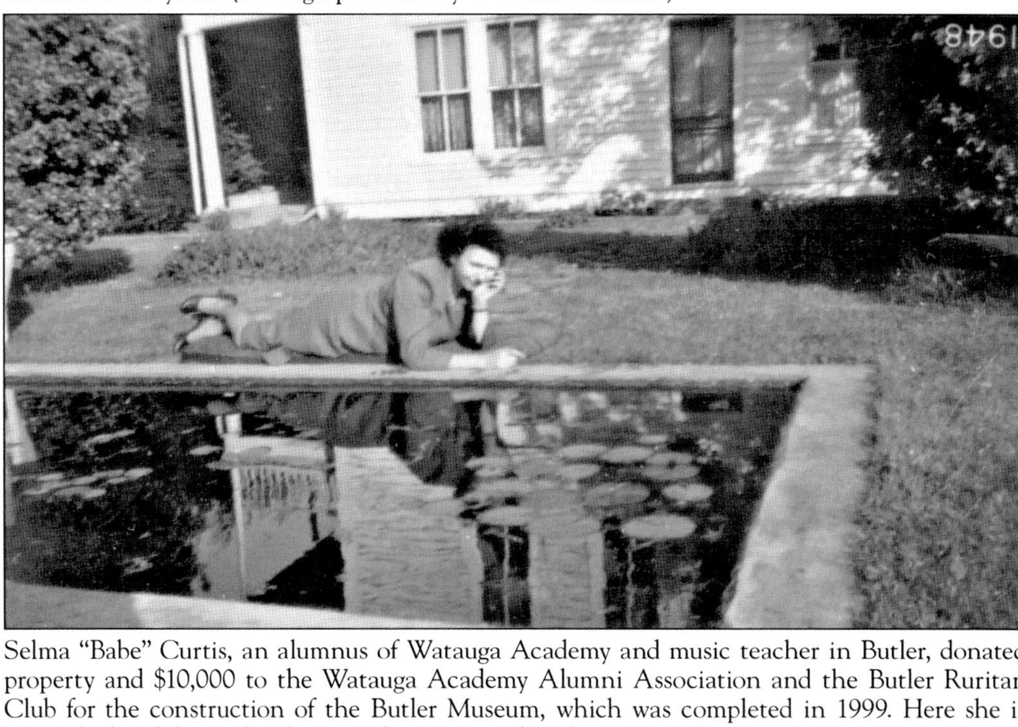

Selma "Babe" Curtis, an alumnus of Watauga Academy and music teacher in Butler, donated property and $10,000 to the Watauga Academy Alumni Association and the Butler Ruritan Club for the construction of the Butler Museum, which was completed in 1999. Here she is sitting by her fish pond. (Photograph courtesy of Butler Museum.)

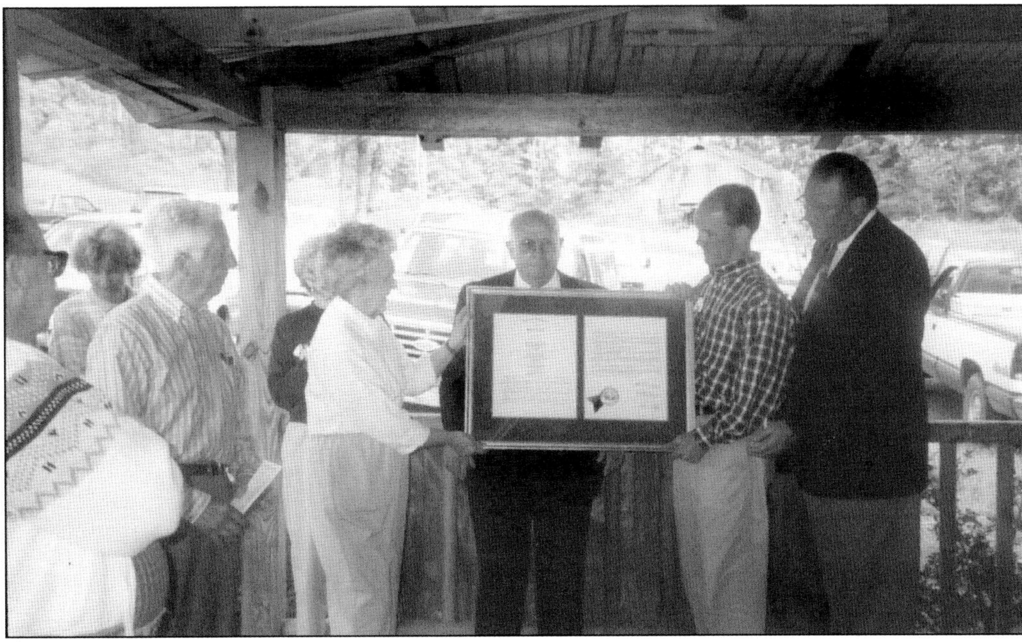

On October 13, 1998, Sen. Rusty Crowe, Rep. Ralph Cole, and Rep. Jason Mumpower presented the Watauga Academy Alumni Association with a framed proclamation commemorating the 50th year since Butler was inundated by Watauga Lake. They also presented a check for $5,000 to go towards the construction of a museum to preserve the memories of Butler and the Watauga Valley. Pictured here from left to right are Patty Adkins, Wayne Eggers, Anna Johnson, Rep. Ralph Cole, Rep. Jason Mumpower, and Sen. Rusty Crowe. (Photograph courtesy of Mary Walker Ward.)

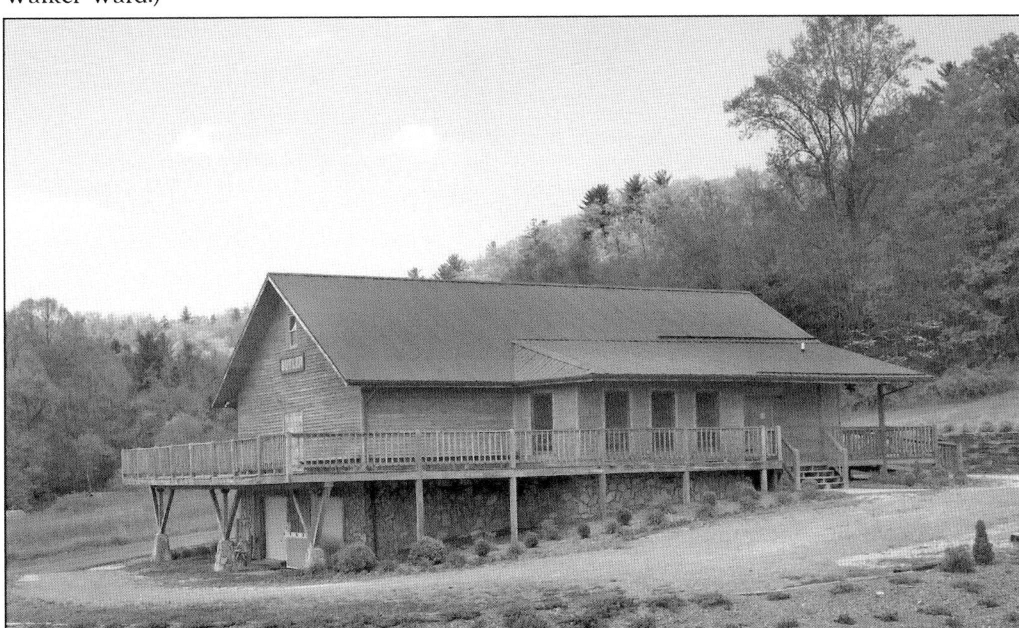

Pictured here is the completed Butler Museum, which now receives more than 1,000 visitors each year. (Photograph courtesy of Michael and Lanette Depew.)

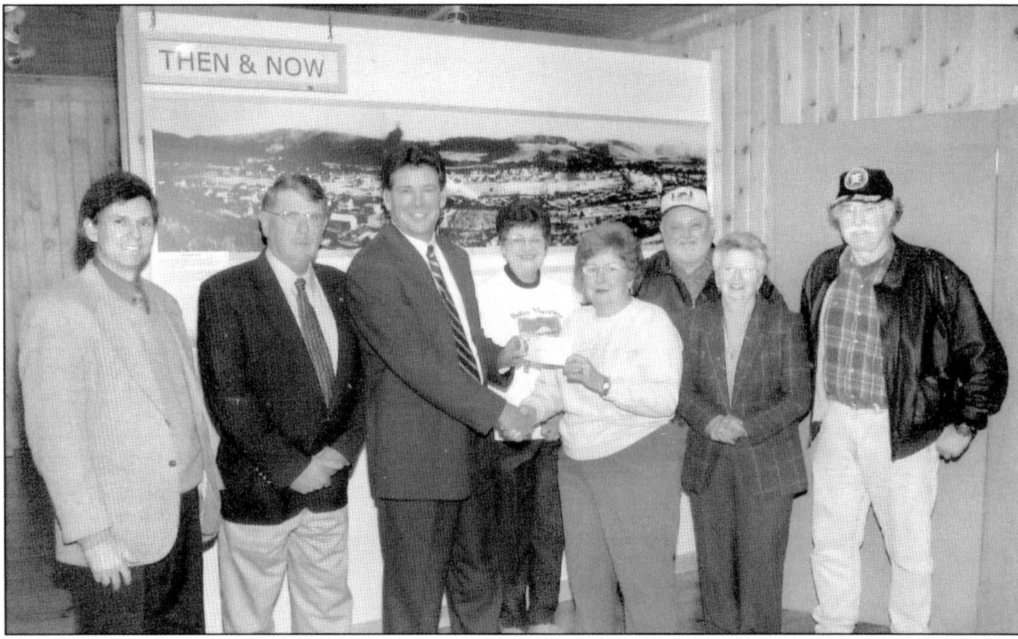

In this image, Tennessee Valley Authority representative Michael Skalf presents a check to the Butler Museum. Pictured from left to right are Joe Thacker, George Lowe, Michael Skalf, Judith Helms, Anna Dugger, Larry Shoun, Mary Ward, and Michael Adkins. (Photograph courtesy of Mary Walker Ward.)

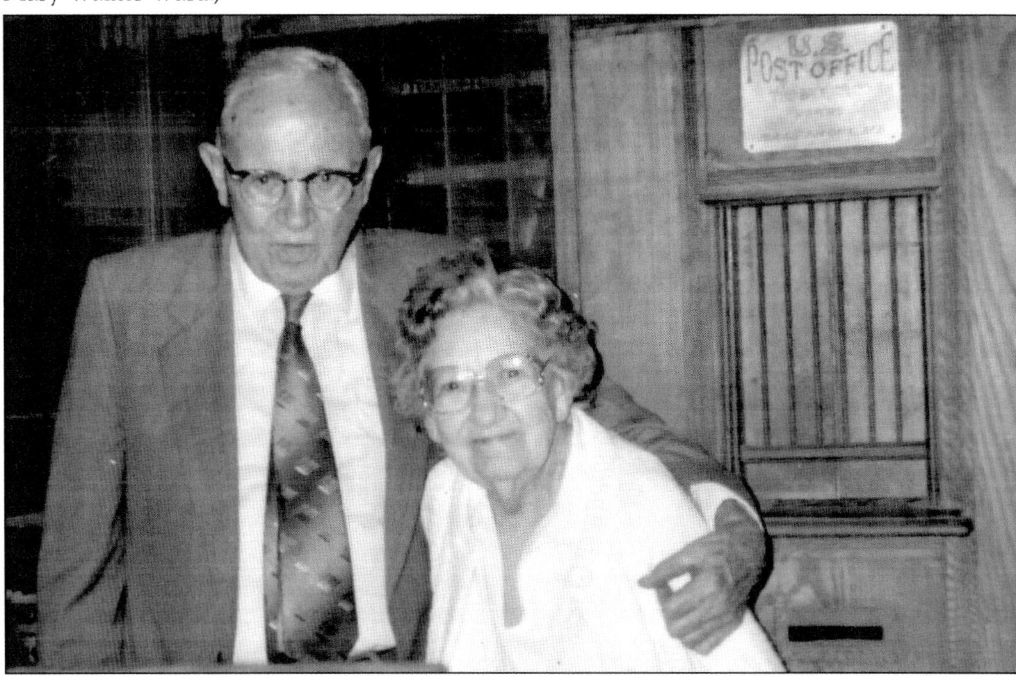

Ted Gregg, a rural mail carrier of Old Butler, and Bess Deal, clerk of Old Butler Post Office, pose together in this photograph taken in the 1990s. The original canceling machine, scale, and mailboxes of Old Butler are on permanent display at the Butler Museum. (Photograph courtesy of Butler Museum.)

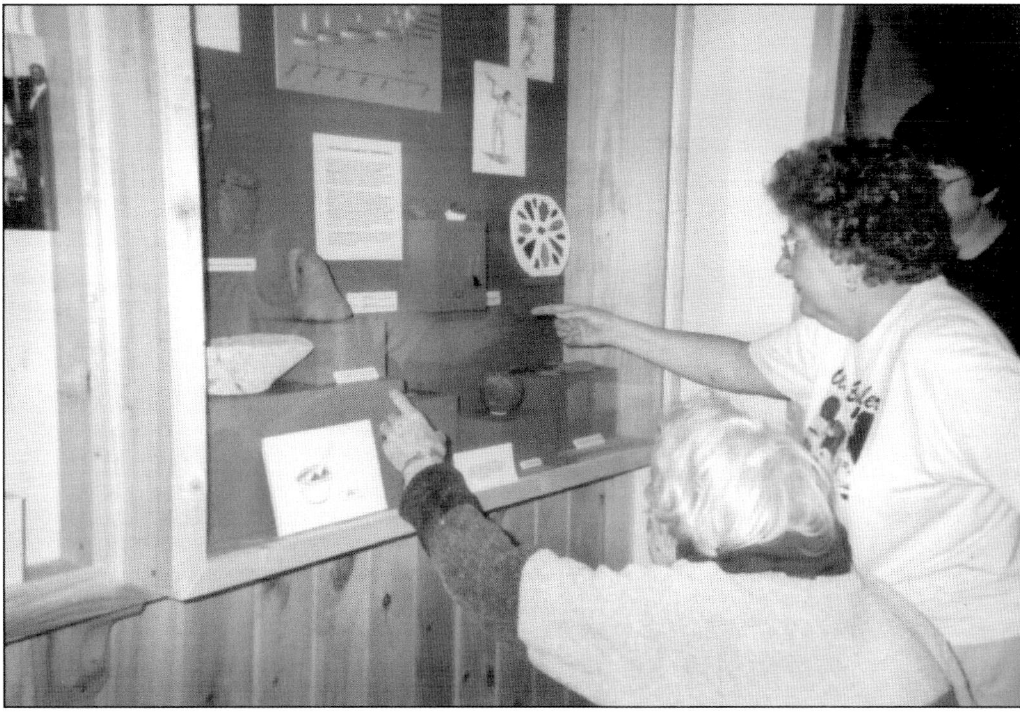

Doris DeVault looks at the exhibit of Indian artifacts her father, Robert M. DeVault Sr., collected from around the Watauga River. (Photograph courtesy of Butler Museum.)

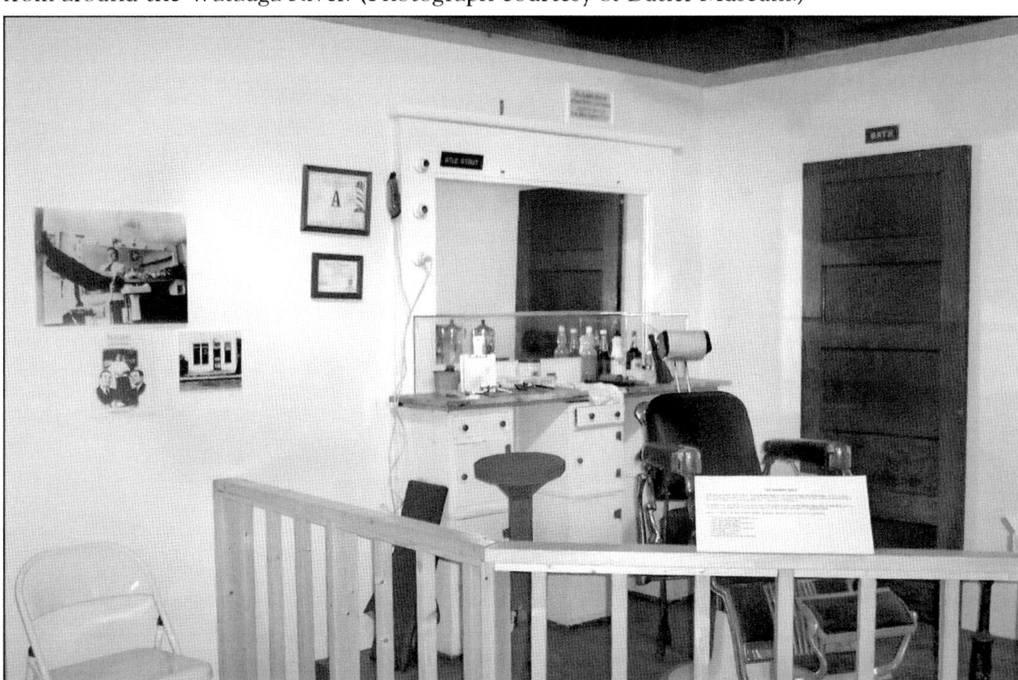

The "Stout Barber Shop" display contains the tools and equipment that Mr. Stout used in his business everyday. (Photograph courtesy of Michael and Lanette Depew.)

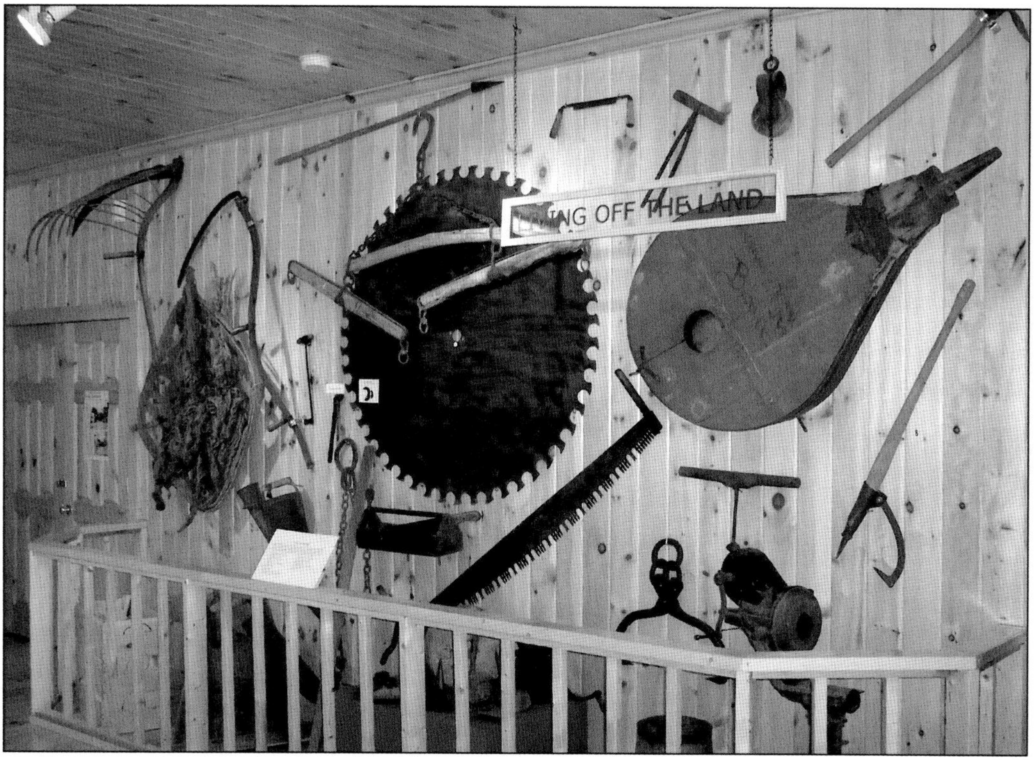

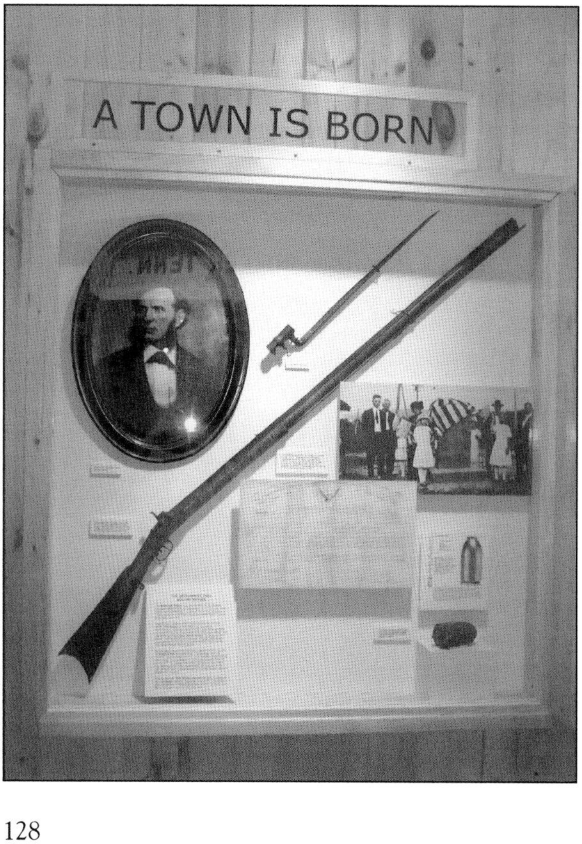

Several artifacts from Butler's early farming and logging days, such as a two-man crosscut saw, logging tongs, and a saw blade from Shoun's sawmill, have been preserved in the "Living off the Land" display at the Butler Museum. (Photograph courtesy of Michael and Lanette Depew.)

The "A Town is Born" display, which includes a photograph of Col. R. R. Butler, a Civil War rifle, and Allen T. Carriger's commission into the Union Army, commemorates the beginning of the town of Butler. (Photograph courtesy of Michael and Lanette Depew.)